FOREWORD

10.22334. A grain of sand.

Variothoughts is sand, not gold. According to Syntalism, our world is built of sand, not gold. The most valuable thing is sand, sand is bread, and gold is salt.

10.22742.

Variothoughts books should be read slowly, chewing every thought carefully. Truth is that which has extension properties, and falsehood is pride, that is, an avid rush.

10.22339.

There are no questions that cannot be answered in the Variothoughts. The Variothoughts is an endless source of inspiration for hearts searching for truth.

10.22266. Nutcracker.

Breaking stereotypes and patterns. Variothoughts is a brain-crushing book, the meaning of which is to achieve the integrity of the mind. First, all beliefs should be destroyed in the dust, and then it will all stick together and enlightenment will come.

10.22554. The living and the dead book.

In the original, Variothoughts is the ideal of the perfection of truth, but the ideal is dead and, therefore, there is no joy in it. To bring back the joy of life to Variothoughts, I decided to salt the bread. Salt-free bread is too sweet. In Variothoughts translations, I threw a couple of spoons of chaos. My act of

1

monstrous vandalism led to the loss of 20% of the meaning, and made the texts very strange and obscure. You will call me a scoundrel and a vandal, but I don't think so ... On the contrary, I believe it made the texts charming, created artificial barriers ... Now, to understand the texts of Variothoughts and find the truth, you have to smash your head and think. Thinking is joyful.

10.15530.

The basis of speech is truth. Cognition of truth should begin with clarifying the meaning of words.

10.16833.

The main feature of Variothoughts is its unprecedented honesty. No censorship of thoughts, absolute freedom of ideas and words.

5.412.

Variothoughts is a book for those who save their time. Ready-made Lego cubes used to put together any ideas and goals. The DNA and RNA of thought.

3.2212. Attainment of truth.

A comprehensive attainment of reality occurs by knowledge's thinning of its tiniest degree of detail.

3.2213.

Unity is hidden in differences. You unite by disuniting. Holding one onto another, the infinitely small becomes the infinitely big.

10.6168.

The Variothoughts is a basic library of DNA of thought, loading it into the brain can solve any problem. Any dreams, any goals will be available to you, thanks to the philosophy of

Syntalism.

3.1971.

Variothoughts implements divergent thought algorithms in order to come again to unity through a multitude. Many grows out of one and one grows again out of many.

3.2211.

Soundness is the ability to dynamically examine things from different perspectives.

3.2216.

Having reached its limit, knowledge transforms into will. What is the limit of knowledge? – Faith.

10.21243. Love of truth.

The meaning of human life is to overcome infinite loneliness and find infinite love.

THE THEORY OF EXISTENCE. DIGITAL UNIVERSE

412.

Hear the hum of the Universe -
countless bees ... the children of it -
turn a wheel of life...

461. [In brevi].

How heavy is the human soul? ... How much weighs 32 GB of information?

4.13. The three-dimensionality of being.

Reality is the borderline between firmness and softness...
Reality is the borderline between water, the shore and atmosphere.

4.84.

You only think that the sky above, in fact, the sky is an abyss, and you walk upside down on the ceiling.

640.

The universe is a super complicated system trying to become aware of itself; in essence it's that very human child, that has a long way to perfection and uniqueness ahead of him.

451. [In brevi]

The world is an illusion: it's exactly the way
...a beholder sees it...
*...another detached beholder sees already a different world.
*

...the world exists only in the mind of the beholder...
*

The more a beholder differs from other people
the more unique world he's able to observe;
....the more perfect a beholder is ...the more perfect
experience he can get from his observation.

7.66.

In the future, people live in the trees. That sounds strange,
but that is practical and convenient.

800. People don't die. They get archived.

4.69.

The truth is, you can take any chaos and, by limiting it to
lines, shape it. Reality is chaos, limited by forms.

9.80.

Viruses and diseases kill the weak and the identical. The essence of this phenomenon is the evolution and natural selection of perfect and unique individuals.

476.

The Universe consists of zeroes and sometimes of ones.

3.173. A one million light years step.

Quantum leap technology allows to fold universe into a
point, a ball... a big ball... a medium sized ball and then unfold it back.

3.461. Perpetual motion.

The wave structure of being is something like a taut bow-string or cocking a spring.

4.85.

You know, raindrops are falling upside down, straight to the ground, and trees grow upside down, boldly hanging right out the abyss of skies.

541. [In brevi]

This world was created imperfect and for the imperfect...
Its aim is to make us perfect.
It's not even a world but a green-house where we should grow...

9.622.

I filled "Variothoughts" with a great number of ideas that could be possibly true under particular conditions. Now, my dear reader, you only need to figure out which circumstances could possibly exist in reality.

10.7093.

Why don't apes evolve into humans? For the same reason that a monkey doesn't turn into a chimpanzee or a gorilla into a baboon. Man is one of the breeds of apes. Previously, there were different breeds of apes of the human species, but then they became extinct or they were assimilated by the main species of homo sapiens.

10.7094. Hobbits.

On Paradise island, the height of a person has decreased to one meter, and the brain has shrunk to 600 grams. When there are no enemies and problems, brains and strength are not needed.

8.33. On the way to Machu Picchu.

The buildings have already smoldered to ashes but their bases are impressive. No earthquakes can damage a really good base.
By the way, recessed walls are also a good element of earthquake-proof constructions. A common example of how to build something to make it last for centuries and start with the base.

8.46.

Boredom is the main engine of civilization, the body, like air, requires surprise.

9.8750.

The three-dimensionality of space is the possibility to set three mutually perpendicular directions, and the fourth one would be impossible to set. That is, there may be only three, not more, mutually exclusive points of view on any phenomenon. All other points of view are superfluous. On the other hand, one or two points of view are not enough to describe a problem either.

9.8661. Parallel worlds.

You are eating an apple in your reality, and someone is munching noisily in a neighboring reality.

9.28. Growth restriction.

The problem is not that there are no answers, the problem is that the questions are over. You are restricted by questions. That's they that draw boundaries that you cannot cross. After all, questions generate answers and answers are a source of growth, expanding the boundaries.

9.69.

I'm tired of illusions, I want to open my eyes and see the real world.

9.3277.

The macro-world and the micro-world are known nowadays, but there should be three, not two, worlds in a three-dimensional space. There should be three, not two, bodies of laws. There exist quantum space, Newtonian space and another, third, space.

9.3276. The dimensionality of space and size.

Essentially the micro-world of atoms and electrons is a bi-dimensional world; a unidimensional one is even lower. Humans live in a three-/four-dimensional world... Going up in terms of size will reveal a seven-/eight-dimensional space.

9.74. Nothing.

It is not the Universe expanding, pushed from within by the energy of the big bang, but it's the void that requires filling, drawing it into itself. Gravity of the darkness attracts light at the speed of darkness. Infinite desire is Nothing, becoming something is the main source of movement in this Universe.

9.79.

The energy conservation law is true, given that time is one of the energy states. The big bang created time in the first place. There was no time before the big bang. As time passes, the process of converting time into energy occurs among other things.

9.5. Scheme of being.

I simplify everything, of course. It's easier to understand this way. Taking the complicated into the simple parts, you will better understand better how it all works. To understand the complicated, it must be taken into the simple parts.

305. Four states of time.

Liquid.
Gaseous.
Solid.
100010110101101101010100101.

3.40. A perpetual motion machine.

The strategy of motivating a person by God is as follows. First, through temptations, man becomes infected with desires and vices, and then, by limiting these desires, he receives the energy necessary for the service of life. Vice need then, to, limiting his, obtain energy.

340.

Only few people descended from extraterrestrials,
others are the descendants of different animals.

356.

Taking into account that matter is information about the form and structure of energy and the fact that the universe expands...there can be drawn a conclusion that the Universe is an artificial intelligence system that endlessly accumulates information and hence expands...

9.48.

Our three-class society has become a fourth-class one. Robots are new emerging class. And that's a strong one, such an adversary in class struggle can easily destroy half of mankind, if not everyone.

4.1.

Yin and Yang are the boundary between the ocean and the sky. The border between past and future. The border is

called now, at the edge of which living fish is the consciousness of man that is afraid to fall in the sky. And there live
birds that catch fish, these are those who are not afraid of the
future and know how to plan it well. In the depths of the
ocean live the demons of the past, sometimes they eat dumb
fish, diving deeper into the past than necessary. And in the
past there is a coral Paradise, it is happiness that always lives
in the past.

44.

There is a couple of perfect utopian communities known
to nature. Hives are perfect female communities.
Anthills are perfect male communities... And of course
no one would deny that extremeties are effective. But
is there truth in them? And who in fact, really needs
these extremeties and what do they finally result into?

9.2.

Question. Does the idea that money can be printed violate
the energy conservation law?

Question. Does the fact that the Universe appeared from a
point and started to expand mean that the energy creation
process is nevertheless going on?

Question. Is the acceleration of the Universe expansion the
source of its energy?

Question. Is the Universe expansion mechanics linked with
the idea of the unity and struggle of opposites and is it a
source of energy?

Question. Money is energy, solar energy is energy, too. Can
we say that the process of printing money is the process of
the solar energy conversion into the money form and it is
linked with the number of people and infrastructure in the
state through derivatives?

Question. Can we say that the faster a society grows, builds and develops, the more money the state can print? In the notion of the link between the amount of the available energy to the society growth acceleration.

Question. If the growth has stopped, can the excess money supply be simply burnt thus making the energy conversion? Energy can exist in various forms. Money is a form of energy. Debts are also a form of energy, a form of money. If the dead debts, accumulated in the banking system, are burnt, it will undoubtedly be the act of energy conversion and release.

241. [In brevi]

The problem of space is that
it is never empty
yes...the density of objects differs
but at least it always contains time.

The human being
one was just a second ago
is already different
from the present self
and every second
this abyss
grows.

250. [In brevi]

It's not that I'm not afraid of death,
I don't really believe in him.

257. [In brevi]

A soul is immortal information,
while a body is just a machine getting
obsolete and unserviceable in time...

266. [In brevi]

Nature intended the human being to be unhappy -
...what for?
To strive after happiness.

310. The world of the wheel.

Everything here is done to make the wheels spin...
Some run to turn these wheels.
Others construct them.
And some motivate others to run.
Giving a carrot and stick motivation.
Apple dream motivation.
A motivation......how many of them...
anyway you'll have to run,
and the ones who don't get it will be remanufactured...

336.

There can't be an old god or a new one...
There can't be old values or new values...
This world is stable enough, it was created
billions of years ago and its bedrock principles
are permanent...

3.78.

The joker loves and is afraid of nothing, and that is the secret
to his might.

3.92.

Interesting. Expansion of the universe after the big bang and
aiming of the order to chaos are evidently the same processes. Creation of the universe is basically destruction of it.

3.93. After the explosion.

Degradation can go on and on and the very fact that our universe exists proves it.

3.94.

Speed is order. The more order, the higher the speed.

3.95.

Miracles originate from the knowledge of mathematics and physics. Consequently, mathematics is the metaphor of God.

3.96.

Scalability is what obeys the laws of progression. Consequently, truth is what lives by the rules of an arithmetic progression. Truth is the growth of life.

675.

An open question also remains about whether information can exist outside this Universe and according to which principles it exists over there...

... about the role of this Universe sees as some information medium...

There's a high probability there are a lot of such mediums. They are created and erased and this initializing code is transmitted between them in a particular way, - resembling the birth of a human soul... It is also born with some basic set of rules and later fulfils its aim and predestination.

677. Axiom of Eggs and Chicken [In brevi]

Here comes an interesting conclusion about one-moment existence of everything and every thing.
Everything - the Universe, time and space exist in a single close moment, inside which there already exists its relative time, like a point in a system of coordinates.

In essence, the understanding of this mechanism helps to understand the concept of an endless point inside which there exists a huge Universe...

The question is only in the effectiveness of codes in the energy of information about its structure...

678. [In brevi]

Beyond the limits of the Universe there are no coordinate constants of this Universe as well as there is no space and no time. Hence the Universe, in essence, exists in a single endless point without time and space.

Perhaps, the very creation of the Universe was connected to some event. Let's call it some informational virus that structured the information about the form of the energy so that it after having produced 4 existing dimensions got an incentive for an endless growth...
like as if it opened ...
like as if there was that very Big bang...

709. [In brevi]

Anything that once was created will live forever. It's only the shell that can be ruined, but the inner essence will exist. What's more the same goes with everything else: people, things and even thoughts. The Universe is information and this information is saving.

7.19. A community of normal people.

There are no normal people. All people are abnormal, but they all want to be normal. Lies are what makes an abnormal person normal. Lies are a mask and an element of socialization that allows people to live in society, pretending they are normal.

7.28.

Space in relation to traffic jams. Time in relation to traffic jam... Now it slows down, now it speeds up.

752. [In brevi]

An absolute point. The point where one moment exists.

When in a single moment there exists all time and when in a single point there exists all space.That's approximately what our Universe looks like from side.

7.55.

Many people are inconstant – one cannot trust them or have faith in them. A man in whom one cannot have faith is, essentially, a sinner.

814. [In brevi]

The main logical motives of history are greediness, vanity, hate, etc. And not some abstract virtues as some extremely pompous fellows like to think.

The base of everything is made of economy, money, power and control over resources and market outlets.

845.

They say there exist three times: the past, the present and the future. But in fact it's only the past that EXISTS, and all actions of a human come down to creating the past...

A person doesn't create the future or the present, but it's only the past that one creates. The future is just imagination, just an intention to create the past. And the present is a process. The process of creating the past. If the past is huge and stable then the present is motion connected with the process of creation.

It conveys the suggestion that the worst crime one can commit is to stop creating the past as this way one will become useless to the world, will stop and become the past oneself...

8.53.

The life expectancy of an individual is directly proportional to the total number of these individuals in the ecosystem. An increase in life expectancy reduces the birth rate. In other words: many people-little oxygen.

New life needs a place in the sun. Big trees leave no chance of small growth, except for parasites. Really... nothing but parasites can survive Near the titans.

8.92. Pleasure and pain.

The thought that extraterrestrial intelligence differs from human, is strange. The essence of intelligence and life in general is pleasure and pain. It's hard to imagine any other essence of life.

9.12.

In particular cases, the energy conservation law does not take into account the "Time" variable. There is an increase in the amount of energy in time in our Universe. The older the Universe, the more energy it has.

9.13.

Time is a form of energy. The course of time is the process of converting energy from its time form into some other states. Therefore, over time, there's less time and more energy converted from it. Human life is, in fact, the process of converting time into other forms of energy and further into matter and information. There is only energy and information about its form in this world. Information is God. Are you asking: where is God? – God is everywhere and everything. And even your thoughts, desires and ideas are God, too.

9.16. Travels are required to think

Travels turn the brain on, connecting it to local information fields. Every new place you visit will give you fresh thoughts that you haven't known before and this will expand the boundaries of your consciousness. The most important id to think, see, watch...

9.17. Video traveler.

Many people want to replace travelling with watching videos about travel, but it is impossible, these things are different. The essence of travelling is not that you see something or relax somewhere, but what you feel. Once in a new information space, you get access to new thoughts and ideas. The more information is available to you, the smarter, more original, more competitive and stronger you will be. Travelling makes a person smarter.

You feel up to 70% of information... What you feel is the result of connecting to the information field.

9.18. Pilgrims.

Travelling like a pilgrim, visiting places of information accumulation and information nodes is very useful from the point of view of the brain and intelligence. Travelling to relax, to lie on the beach or to get drunk, visiting the same places again and again is useless from the point of view of the brain.

9.20.

Accessing information through books and videos is difficult. To understand something, it's necessary to feel it. And a person feels up to 70% of information. To feel the information, you need to be physically in the place where the information is located (information node) or next to the person who is the carrier of this information.

Summary: to achieve the power of mind and turn the brain on, you should travel more, communicate with clever people and find a good mentors and teachers.

9.21.

Taking into consideration that up to 70% of information is transmitted through senses, you do not even need to speak or read (it works both ways, though). You can just stand or walk around the source of information and your head will be filled with original thoughts and ideas. Enlightenment and comprehension will come upon your brain like waves.

9.25.

The ant that can't cross a drawn line is like the person who has a certain set of thoughts in his head, outside which he can't basically think or act.

The point is that crossing this line is like crossing the border and plunging into Nothing, there is simply nothing there and there is absolutely nothing to do there, but expand your boundaries. To expand your boundaries you need to travel and communicate with clever people. You can also read, but reading is not a panacea, it only provides 20% of the information, the rest should be felt. And to understand, you need to feel and to do this, you need to be close to the source of information and feel it.

9.26. Information comprehension.

Information can and should be comprehended, but a lot depends on the strength of the intellect (logic, dialectics) and the power of the subconscious (the inner It). The subconscious refuses to expand its boundaries as it is dangerous. Everything new and unknown is outside the scope of the past experience and is therefore rejected by the It. Rejection of information occurs at the emotional-hormonal level: being bored, disliking, feeling pain, discomfort, aggressive disagreement. As a consequence, re-comprehension of information is available to very few people. It is difficult to get rid of the It's dictate. It's possible to solve the problem by coming near the source of information. In this case, the outer social It is turned on thus turning off the inner It, forcing it to perceive information. In addition, direct contact with information allows to feel and understand it, having received previously unavailable 70% of information.

9.27. Growing from the inside, you grow from the outside.

There is no need to cross the line, there is Nothing and Emptiness behind it, but it's necessary to expand the boundaries using Nothing as the free space for growth. That is, it is necessary to grow from the inside, thereby growing from the outside.

9.30. Meaning of an ant's life.

Do you also think the meaning of life is to eat, to go to the toilet, have sex and build anthills, filling them with sticks and leaves?

9.31. To understand a person, listen to his silence.

You don't have to say anything to say something. The words are only 20% of the information, the rest is felt by your interlocutors. Moreover, silence is more sincere than words, it is a kind of telepathy, it is more difficult to deceive feelings than ears. The same applies to you - listening to stillness and silence, in the pauses between the words, you will hear and feel much more than words can say even theoretically.

9.32.

To hear and understand the city, you need to wander through it alone for about eight hours, listening to its breath and then you will hear all its thoughts, its soul, its ideas.

9.33. Information nodes and objects.

Historically significant places and artifacts, scientific towns and universities, natural energy objects, beautiful places, huge megacities, places where great people lived and died.

Once physically in such a place, a person can access the local information field and download new information or comprehend and rethink old information.

You can also access information through a living person. The more perfect and more unique a person is, the more information you can gain, even just standing beside him or touching him...

One connection session is like dinner at a restaurant. The menu is large, but you can only try a couple of dishes. To extend the pleasure and try the rest of the dishes, it is necessary to do this procedure many times and at every opportunity.

9.38. Life and death are the same.

Peace is death, movement is life, but peace is emptiness and free space. Free space is required for movement to occur. On the other hand, life is rotation, movement in a circle. And sometimes, life is a pendulum whose oscillation amplitude determines the limits of its movement. So you don't even have to have free space to move. Movement is life, because movement is the process of energy conversion: a dynamo, an electric generator... You move to live, you live because you move. Moving, you get energy, you need energy to live.

9.39. Pendulum on the edge of Nothing.

Movement is necessary for life, the bigger the movement, the more energy is available. But when there is little space, it is difficult to move. If you get close to the edge and cross the line, you'll end up in Nothing. Nothing is where there is a lot of free space - the space where you can move a lot thus getting a lot of energy (money). But beware: Nothing is addictive and can easily devour you. STAY close to the border and come back quickly. Maybe the pendulum strategy will help you.

9.40.

Motion is a source of energy. Think of an electrical generator. In order to produce energy, you need to spin and move, turning time into energy.

9.42. Potential motor power.

To think, you need to think. Thought is information that is generated by the movement of thought. Though is what manages energy, structuring it into matter. The bigger your thoughts are, the more energy will be potentially available to you when you start moving. The power of thought is the potential power of a power plant that can begin to produce energy (money) if you start moving. But to start moving, you need energy - fuel, wind, water, sun, coal, etc. In other words, your intelligence is the potential power of the system, movement is the process of energy conversion and you also need fuel.

9.47.

They create the world not for people and that's great problem, as to create a new world, the old world, the world of people will have to be destroyed and then people will die out like mammoth.

9.51.

God is long dead. After he created man, man killed him. The man needed free space to live. The old gods took up too much space and consumed too much energy.

9.52. Evolution of human into a radio-controlled ant.

The technology of an electronic assistant based on artificial intelligence, which will help human to remember information and suggest something, will most likely lead to total degradation of the human brain, which will quickly destroy humanity. Now, when it's not necessary to think, the brains will be reduced even more and human will turn into an insect. The assistant will manage human through a radio channel, and human, as it befits an ant, will crawl and collect energy, dragging it in the anthill. The more energy the ant gets, the happier it will be.

9.59.

If you have a question, there's an answer to it…
There are no unanswered questions.
The main thing is to have a question, the rest is not important.

9.63. Multidimensional world.

Everything has at least three sides, because we live in a three-dimensional world. But usually there are four or more sides, because there is still time and that makes the world at least four-dimensional. Illusions make the world multidimensional.

9.64.

Theoretically, a qubit should have not two, but three or more states in quantum physics. We can say that the dynamics of quantum states stretched over time is information.

9.67. The last God alive.

Turning the society from the three-class one into the two-class one is degradation, as three-phase systems are more energy-efficient than two-phase ones. Moreover, class contradictions were an important source of progress, if they are reduced, the society will start degrading. Furthermore, when the elites replace the lower classes by robots and artificial intelligence systems, the issue of what to do with the majority of people will arise.

The answer will be evident – to immerse them into illusive

dream to make them perish in sweet pleasures, like a fly in the syrup. This decision will turn all the social elevators and ladders off. The renewal of blood and thought among elites will stop. That is, interclass motion will disappear along with classes. And motion, as we remember, was the basis of life and an energy access mechanism. Anyway, everything will end and perish rather quickly. In a couple of thousands years the last God alive will create a new Universe and, having erased his memory, will self-destroy.

9.68.

Creating robots, they actually want to become gods. Having created an artificial mind, they will certainly succeed in it. The only thing - this has already happened in mythology. Wallowing in pride, Devil once declared himself equal to God, for which he was instantly sent to hell for eternal suffering and torment.

9.71. Free man.

It often happens that luck finds completely inarticulate people who need nothing, who do not have dreams and their own ambitions. This is due to the fact that these people are potentially useful, but free. Therefore, any external Goal can easily find them and use them for their own Goals. After all, if these people had their own goals, they would have to be repulsed, they would have to fight for them, but this way they are free and this makes them easily accessible. Their emptiness creates a vacuum that attracts something that could fill them. In this case, it is strategically correct to choose the best offers.

9.72. To find the new, you should get rid of the old.

Emptiness seeks to be filled. If you need something new, you need to make space by getting rid of the old. Throw away everything old, and the new one will come to you. This technique can be used, for example, to upgrade your wardrobe, or to find a new love or a new job...

9.75. Mutual love.

Nothing wants to be filled and energy is looking for Nothing to become matter having filled it.

9.76. Unique person.

It is especially dangerous to take someone else's place. When you take someone else's place, nature will decide to get rid of you very soon, sending you to diseases or misfortunes. To be in your place, you need to cultivate uniqueness in yourself. When you are unique, there will be no one to take your place, therefore, nature will have to take care of you.

9.77. Healthy child.

The farther genetically the parents are from each other, the more unique will be the genetic code of the child. The more unique the child, the harder it is for diseases to eat him. The more unique a person is, the better his health is.

9.84.

The more people there are, the more similar to each other they are and the easier it is for diseases or other troubles to eat them.

9.85. Something against which there is no weapon yet.

The essence of uniqueness and novelty is that you are so harder to eat and harder to fend off.

9.88. Mechanics of protection against immunity.

The mechanism of constant DNA mixing during human reproduction is needed either to protect against viruses and bacteria, or because the person is a virus, and he needs to constantly change to protect himself from the immune bodies of the environment where he parasitizes. By the way, these immune bodies, apparently, are called viruses, fungi and bacteria.

9.98.

Man is a hoofed animal. You don't believe me? Look at your shoes...

9.99. By the way...

There are much more pigs and hens than people on our planet.

1007. [In brevi]

The truth and verity lie in the balance of chaos and order. But it's not the chaos and order that various religions present. It's not evil and good but rather some system rules and approaches towards distributing energy.

Chaos demands variety and limiting energy for one creature... It creates thousands of small flowers and some of them will become perfect and unique...
And order is known for not limiting this perfect and unique flower in resources as it spends a lot of energy so that the flower will attain perfection sooner or later.

1096.

The Universe is a big self-teaching system, the goal of existence of which is about betterment and creation of new forms of information. Thus, there are mechanisms which fulfil this goal.

1098. The principles of systemic rationalism:

Everything leading to the creation of new forms of information is rational.

This includes:

The spread and transfer of information and energy are rational.

The system's existence is rational.

Processes related to the creation and showing signs of novelty (uniqueness) and perfection are rational.

Consumption of a lesser amount of energy to achieve the same result is rational. The energy saving principle is so too.

Destruction of essences showing no signs of perfection or novelty is rational. The same goes for making room for the creation of new things.

Both direct and indirect participation in processes related to creation or destruction is rational. The help and support of individuals who create, destroy, spread or transmit information and energy are rational.

1099.

Eastern doctrines speak about the harmony of inner and outer, about the release of wishes as the sources of pain and energy loss, and thus they make idiots and vegetables of normal people. And this is clear. These doctrines appeared in highly populated regions, where people had no possibility for normal life. The local elite desperately needed some ways to release the tension in suppressed people. But in general, these philosophical doctrines have a lot of common with Syntalism. Syntalism agrees with the statement that the unrealized wishes bring pain. It agrees that there is infinite energy source inside a human and that the inner side grows together wit the outer. But as Syntalism agrees in details, it does not agree with the whole thing — the release of wishes and targets. The Universe has a purpose, and all human life is based on this purpose. Syntalism speaks about correct targeting, about the sources of power and happiness. In understanding of happiness it goes far beyond the limits reached by eastern doctrines. In particular, it says that a human being will reach maximum happiness if in his life he follows the targets and meaning of the Universe, and the paradise for him will be here on Earth, not somewhere in heaven. Syntalism defines happiness as a natural response to rational activity...

1106.

In this world, everything is information or information about the form and structure of energy. Everything in this world is a system and everything consists of systems and is part of the system. This is why Syntalism perceives everything from a uniform point of view and establishes uniform control principles for everything in this world instead of dividing nature into subjective and objective realities.

1124. Life and youth.

It's not just that Vampires live in a cold dark crypt and come out only at night. This way of life is associated with the desire to live longer than necessary. There are factors that can slow down aging and keep the body from changing for longer.

All these Vampires, Snow Queens, sages and young princesses in the mountains and high towers – all these people retain freshness for a long time.

It is believed that all "systems" on Earth exist under the same laws and principles. Remember, to keep the Apple longer, it is necessary to tear it a little more immature and store it in a dark, dry, cold place, it is desirable to limit the flow of oxygen. Oxygen in this case is the basis of oxidative chemical reactions, which are excellent in heat and light, and lack of oxygen, cold and darkness such reactions can significantly slow down.

You can also save the "Apples" by drying them or canned in syrups and brines, you can even alcohol.

Youth and life prolong the following factors:

- Highlands (and other places where there is a lack of oxygen).

- Cool and even cold (highlands, North, air conditioning).

- Do not sunbathe, avoid direct sunlight, wear closed clothes.

- Dryness and limitation of water (including fat). Dryness is needed in the body and around.

- We need to slow down the metabolism. There is should rarely, but very refined and a bit.

- Various preservative impregnations and tinctures (there is a great risk of killing vital organs, however, if you treat it wisely, there may be a benefit). Retains not only dryness, but also Vice versa-full immersion in solutions and high humidity-are an excellent preservative.

Additional factors.

- Well oiled iron tools and parts wear less and are subject to rust. So people have to 30% of their energy be obtained from oils and fats, preferably fresh and without any unnecessary chemical treatments.

- What moves moderately, less rust. Rust and oxidation is a disease of metals that causes their destruction.

"Salt is a reasonably good conservative. How to consume salt is a controversial issue, but all that lives in salt water and closer to the ocean, for a person is quite useful. People who prefer marine products are well preserved.

"Other things being equal, it's better to eat the food that

took longer to ripen." Beef is healthier chicken, a healthier sturgeon, carp, late varieties of apples are better early varieties.

- It is necessary to eat only fresh and Mature food. Eating stale, immature, overripe or old food is harmful to humans.

1419.

It appeared in my dreams,
that our world ...
is inherently a game
of some people long extinct,
who lost their bodies...

Food for their thought -
the place, where they could
... play into life,
to remember it the way,
it used to be some time...

1595. Source of pleasure.

Modern life made it possible for a woman to choose a man not only because of money but also because of his soul and other traits.

Nowadays a woman can make money on her own and can protect anyone. So now it's not only money and power that are important about a man. Now a woman may choose, for instance, a soulmate... or other traits in a man. For instance, kindness, tenderness, care, the ability to be a perfect lover... It can be said that nowadays a woman needs a man for pleasure...

1886.

A three-level person. With the face hidden under various masks. The masks hide a monster. The monster guards the soul made for love and light. Hence, masks hide the monster who guards a tender kind soul.

The monster is necessary so that no one would ever smear a fragile clean soul. And the masks are necessary so that no one would panic at the sight of the monster...

1946.

The concept of a bored God lies in the idea that the world was created by God who felt bored. The goal of this world's existence is in giving his boredom the brush-off. We, people should create something to make him wonder and heat his imagination. He's in acute need of novelty and perfection, beauty and astonishment...

Hence, the most of wealth in the world will go to those people who are the best at giving the Supreme Being's boredom the brush-off or at delighting his well-seen eye with beauty and perfection.

Even wars and destructive actions are accepted by this God (but without any fanaticism), as they bring diversity to his boring existence and clear the space for the new.
This God is also into humor and jokes. He likes to make fun of people and people to make fun of each other. He's the main joker. Every good joke makes him laugh and feel nice.
He likes human passions. Passions are his special forte.
Life's an endless soap opera full of feelings and emotions.
Perhaps our God is not even alone. What if he's got friends whom he invites to show what happens to us here on earth.

2121. The concept of inkblot.

The universe resembles an inkblot...
Cosmic chaos... absolute novelty and eventuality...

But we remember that one really wishes for what one doesn't have. Absolute chaos wants absolute order.
The Universe likes the beauty of perfection most of all.

It's a sum of uniqueness typical of nature from birth and perfect beauty that order creates...
Unique perfect beauty is probably the most precious thing in the whole Universe.

The unique tends to become the perfect...

Rapidly spreading to all directions, the inkblot of the Universe which is full of uniqueness and chaos, above anything else craves to become a perfect immaculate dot back again.

Perhaps, some day in the future it will really happen. The universe will shrink back into a dot and will gain its original perfection. But perfection cannot be stable in this Universe, and that's why as soon as the Universe becomes a perfect dot, it will soon be reborn and it will cause a new Big Bang. And the chaos will start again.

2272. A spiral of evolution.

The rise and fall of humankind are two different phenomena.

1% of people evolve and become «Gods».
99% evolve and "turn into bees, ants, aphides, wasps, worms and flies»...

At first it will cause a rise but in time, Gods who don't face difficulties and competition, will fall. The rise will end...and a long night will come. Later, in a million years, there will be a new rise and a new spiral turn.

2296. Immoral robot.

Machine life and people-machines, living in it as robots. An immoral robot - I knew one. They forgot to program him for high moral standards. Not that he suffered much, but still.

2364. A person lacking contrast.

A person with syntonic personality is the one who got a contrast of two equal halves, namely happiness and unhappiness. And it's one half that makes the other one so wonderful and necessary.

Being half-plunged into peace, one lives in active action with the other half. As it's only action that lets one feel the joy of peace...

Enjoying pleasures and joy with one half, a person feels sadness and light pain with the other half. And even the wisdom of this person becomes true only in contrast with this person's stupidity.

A person who lost the balance and get deprived of the contrast
doesn't learn the truth in anything and is full of deception.
– I call it « a person lacking contrast».

2394. Murphy's fifth law implementation in dynamical astronomy.

The Universe craves for chaos (the inkblot concept). The forces of order are aimed at creating perfection and some structure. Hence, there are two types of forces which are opposed to each other.

When it comes to human life, the force of order is represented by human beings (while wild nature has its own laws).

Thus, any unruly thing that a person leaves unattended, will definitely go wrong: slowly at first, gradually gaining speed until it gains critical mass after which there will be a blow and the whole thing will turn into chaos.

2453. The civilization of the 27th generation with the artificial intelligence system of 8.0 version.

And now let's imagine our whole life as a computer game. As some economic-military strategy. Since the beginning of time and up to some distant future there's a task to build a civilization. Inventions, new towns, economical development, wars, building maintenance. And God is our gamer. The game is perfect and powerful with all its 4D technologies, characters (artificial intelligence system). Everything's adultly made! Even multiplayer game mode is available. Thus, it's possible to play online. So that our God can be joined by a dozen of friends to get all jollies. It's all on a cosmic scale.

But still, our existence can be stopped any second. The computer may break or run out of memory or electricity may be cut off. Or our God (in case it's a young boy) may get bored and go to sleep.
So, being the artificial intelligence system of 8.0 version computer game characters what are we supposed to do? We're nothing but information. We're the perfect algorithms of a perfect computer system.
The first thing that comes to mind is to make our scientists invent something that would let us into the world where our God lives. At first sight, it's generally impossible as we're intangible while the God's world is full of tangible objects. But!!! It's just a supposition. So what if that world is intangible as well? What if that world and our little fun-loving God is also a character of someone's game with the artificial intelligence system of 16.0 version?

By the way, what do we need this boy's world for? I suggest that we should recode our code into viruses and try to control our computer and all the games on it. That's for sure-thousands of parallel worlds. And what if we manage to get into the net and get an access to millions of parallel worlds. Generally, if we manage to become viruses and control the computer system of the outer world where our young God lives, we'll be quite able to control the whole outer world, ... all parallel worlds and maybe will even manage to get one level higher...

Quite a game ...Isnt'it???!!!!

2459. The rules of the game may be very idiotic.

According to the belief that the whole world is a game, our God has to play by the rules as well. But it's another matter that the rules may include bonuses and superpower, miracles and new lives. Anyway, if this is a game, we'd better learn the rules.

2462.

In the process of making the impossible a reality miracles happen more often than usual.

2463. The Farm 8.0 project. The system of artificial intelligence 13.0.

We live in the reality simulator called The Farm, generation 8.0.The essence of its existence is in growing systems of artificial intelligence (human souls) suitable for further exploitation as part of various scientific and research projects as well as of service, construction and military systems. Well-balanced souls that have become harmonious and perfect are considered to be first class and suitable for further use. While other souls that have sunk into vice and failed to achieve balance, get processed into raw materials.

An additional extra product.

The Farm 8.0 project also specializes in producing works of art- such as literature, music, cinema, architecture, sculpture, visual art and other arts, humor and entertainment. It's the souls of the special high value that managed to succeed in producing the additional product and becoming perfect, unique and really useful at doing it. Souls of this kind are of ultimate importance and really rare. Producing the products connected with arts and entertainment brings the Farm fifty per cent of income.

A perspective additional product.

In prospect, as the Farm project develops, there's a plan to get extra income from advanced scientific and applied technologies. As of today, there exists the selection and evolutional development of the "luminous souls" capable of scientific activities. It's already possible to mention some powerful success and this module is planned to be put into practice in years to come.
However, those very souls suitable for being used in scientific research activities, are already being successfully grown and they show great results in being exploited. The share of this field's profits for today is 13%.

2608. It is not given to a man to see the truth.

Let's imagine that every person lives in some abstract world. And there is some information circulating in this world. And this information is broadcasted in the Internet and mass media and people bandy this information. Many people automatically believe everything they say. While some people are nihilistic as they automatically doubt everything others say and misrepresent information.

And now imagine that there is no truth in this world. No truth at all. Some say it's white and believe it while others automatically twist it and try to prove it's black. Thus, people argue whether it's white, grey or black, or even green. And people twist, change and misrepresent their matter of dispute. While in fact it's «transparent with a shift of tone 1980». And this world simply doesn't have such a notion as «transparent with a shift of tone 1980». And human eyes and brains are unable to realize this notion.

2611. An optimally set brain.

A brain set to see meaning in everything. A brain set to notice common sense and balance. A brain able to see both trees and a forest, form and content. A brain that understands dialectics and sees the nature of things. A brain set to fulfil tasks. An observing brain... A thinking brain...

The ability to think is a great joy and happiness. Eternal joy of understanding. Eternal power of stars.

2657. A new list of sins.

- Something done by halves or left unfinished.
- Something done badly. Reject workpieces.
- Implementation of something that is not new, perfect, beautiful, unique or useful.
- Inefficient use of energy.
- Ineffective use of time or wasting time.
- Overproduction. Production of something useless or excessive.
- The disturbance of balance or sense of moderation at anything.
- A refusal to fulfil any necessary tasks and predestined things.
- Any acts of supporting or taking part in anything that doesn't bring novelty or perfection.
- The state of being generally useless during one's creating of something to bring novelty and perfection.

/ This should be explained. It's possible to be useful in different ways. It's possible to help someone or do it yourself, or be a grateful user, or clean the space and provide materials, etc. But when a person doesn't want to do anything of the mentioned, this person will burn in hell from any moment of lifetime and long after.../

2699.

The four basic matters of our world are perfection, novelty, use and beauty.

2700.

Every Goal needs a Doer in order to produce a Result.
A result is a child of the Goal which basic characteristics are use, perfection, beauty and novelty. Such a child will be liked by the providence. And the providence will help this child come to this world to the best of its limited ability.

2701. A universal constructive scheme.

The providence needs a child - a Result that has some characteristics (as we have already mentioned, perfection is among them). And the providence chooses the best spouses for it- the Doer and his wife the Goal. The doer should adore his wife and simply crave for her. Also, the Doer and the Goal should have a nice genetic background (i.e. possess perfection and other qualities). The Doer should be a good husband. He should be energetic, smart and able to keep his wife during pregnancy.
Thus, if a couple suits the providence and the Result of this couple's joint working seems promising, the providence may somehow support this couple. It may help with money or moral support, bring good luck or even wonders to their life, especially if the future child seems to be really nice.

P.S. By the way, it even works with real couples who dream of becoming parents. If everything about a couple is nice and fine, then this couple will be on a real hot streak.

It works with everything, be it starting a family, building a house, running business or creating works of art. It's a universal constructive thing.

2850. Evil truth...

One who lives only in joy and pleasure, becomes a disgusting obese soul full of sins... It's necessary to suffer to remain human. Joy and suffering should take turns to keep the balance.

They say that freedom is a necessity and that all people are born free. But it's not true. A person needs only joy and pleasure to feel happy, and pain and suffering just for the contrast. While freedom is just a mask of vanity and the desire to be in power: at first over oneself and then over others.

The fight of good and evil doesn't exist. There's just some mechanism that makes some check-and-balance system work so that to make people fulfil tasks that providence prepared for them. A person who does something good, gets sweet carrot. But sweet carrot makes this person fat so that it's stick and not carrot that is necessary for making this person fit again. By the way, one who doesn't do what one should, becomes treated with the stick approach as well. Nothing personal, just celestial mechanics.

2988. The art of living.

Realization of any plan may end in anything. The result may be unpredictable. But the most important thing is that no matter how it ends, there will always be both advantages and disadvantages. The main art here is about deriving benefit even from the unpleasant results.

3005. The ecosystem of pluses and minuses. [In brevi]

Every good always has a bad and vice versa.
*

A good deed is when we create pluses to compensate for someone's minuses. A bad deed is when we create minuses to compensate for someone's pluses.

3006. Perpetual motion machine.

Order tends to organise chaos and chain it.
But chaos craves for freedom. And freedom is a sensible awareness for chaos. This is that very perpetual motion that makes the sun shine. Chaos is endlessly various as it's a god

of novelty. It figures out many ways to escape order. While order is aimed at perfection. Chains, cages and walls constructed by order become more and more perfect.

I can even imagine that sly order uses the desire of chaos for escape as a hamster in a wheel to spin it and produce electricity.

In the great scheme of the Universe we see two things. On the one hand, stars are formed while on the other hand, they are destroyed, but... it's not that important. But the most important thing is the very black inkblot of the Universe, that very chaos that expands endlessly. It never stops. The Universe expands and its edges are stormy with chaos while the centre is filled with old galaxies full of stars. We may only suppose that Order is some kind of virus that needs chaos to feed on it and use it as a material of construction. Perhaps, Order itself can get its GOAL only after having got the chaos DNA that has novelty and energy.

So who will win in this struggle? - Nobody. The essence of this struggle is in the moving forward and if something wins, the moving will stop and it will be the end of the world.

3007. A cup of invigorating coffee with elements of Metaphysics. [In brevi]

Chaos brings novelty and the source of endless energy - the energy of the big bang. Order tends to get this novelty and energy in order to get its Goal by using them and create something necessary that possesses novelty and perfection. Perhaps, this something is called Beauty or maybe it's just a result that Order needs.
*

«I saw a man in every woman» /// logical guess №1

But that's when I thought of another thing. A man tends to be with a woman. Even though he knows that she will take his freedom. All this material world craves for beauty. Beauty, i.e. perfection is the most precious thing in the world.

*

«Every woman is a better version of a man. Being close to a woman, a man tries to find himself in a better version» /// logical guess №2

3011. Perfection is something that is better.

Beauty and perfection need uniqueness because it's impossible to choose something better from alike things. Uniqueness lets everything be the best in its own way.

3071.

We live in a visionary world one third of which consists of reality.

3111. It's better to be in joy than in sorrow.

The only difference of a positive-minded Universe from a negative one is the good mood of its inhabitants.

3116. An all-embracing definition of use.

Something that helps desires and needs translate into reality is really useful.

3134. Physics.

When perfection dies, it releases a great amount of energy that will feed everything new that still tries to become perfect.

3.161.

The point is, boundaries can be pushed. Boundary signs usually begin to be installed on the last mile, which takes 80% of the way.

3.174. Virus of beauty.

Beauty and the truth are the same. A scalable system that is able to enslave and control energy.

3.176.

Any action entails both a good consequence and a bad one at the same time, but it depends only on you which one you'll ignore and which one you'll support.

3246. The balance.

The Universe of existence and nonexistence.

3309. The wheel of life.

A person is a squirrel spinning the wheel of life. And various human passions are just like stick and carrot helping the squirrel do its job.

3.362. The man who destroyed our world.

The journey into the past is the process of re-creating the past and turning it into the present. Going into the past, the traveler destroys the present and creates a new present, which naively believes the past. Such a traveler will never be able to return to his reality again, because he destroyed it. Whoever invents a time machine and attempts to travel to the past or the future will destroy our world by erasing it. - Are parallel processes possible? "Yes, but this man won't know how to use them.

3.390.

They say stars create time, and black holes recycle it.

3.417.

A circle is a harmonic fusion of a square and a triangle. If the square starts to rotate, it will turn into a circle, the triangle movement also turns into a circle. The circle is a symbol of movement, a symbol of time and harmony.

3.420.

The faster we move through space, the slower we move through time. What does that say? - That space and time are two interrelated entities, obviously two sides of the same coin.
"What does speed have to do with it?" - Obviously, accelerating, we begin to catch up with time, equalizing our speeds.

"What does that give us?" – First, it gives us an understanding that time is a state of motion of the flow of energy. The material is that which has lagged behind, that which has slowed down, perhaps has died. Things die when they stop and time overtakes them forever. Secondly, the speed of galaxies, stars and planets, forming a single vector, creates a local time constant, which, obviously, in different galaxies will be different.

- Does this conclusion have any practical value? - Speed up and you'll outrun other people's time. The greater the order, the greater the speed. Things die when they stop. Light carries mass. What does that mean?" This means that the slower you move, the more energy is taken away from you, the faster, the more energy you have left and even added.

3.427. The speed of time is the reference speed.

Time is not even speed, but the difference (Delta) between your speed and the speed of light (or some other reference speed).

3.429.

There's a huge correlation between rotation and time. The rotation implements the wave model phase and protivogaz. Rotating object that accelerates time, accumulating his energy, then slows down, transforming this energy into other forms

3.437. Slow down.

It just seems like the order increases the speed, in fact, it reduces it, but gives it a vector. The speed of chaos is higher than order, but it is random and has no purpose. Order is the purpose and inhibition of chaos.

3.438.

The lower the speed, the faster time flows, that is, time increases. Deceleration increases, not decreases, the Delta grows. Light is something that dies instantly and is born instantly, that is the nature of waves. Those who have slowed down live longer, but their death is no less prolonged.

3.439.

Approaching a living object to the speed of light will not so much create a temporary anomaly as destroy the object. The cycle of life and death will be shortened. Living matter cannot travel at the speed of light, but information can.

3.440.

Chaos is the highest form of order. Chaos adult corporado. A more simplified and delayed form of chaos is called order. Order does not collapse into chaos, it grows into it. Chaos, on the contrary, is destroyed in order.

3.442.

The essence of the wave is the spring and the accumulation of kinetic energy of time. In the negative half-wave time, like a spring, is inhibited and energy accumulates due to destruction, in the positive half-wave stored energy creates an explosive effect.

3.444.

The nature of time has the same nature of dualism as light.

Time is both a particle of matter and a radiation of energy. An ultra-high-frequency oscillation, where every moment our world is destroyed and created anew.

3.448.

Light is something very small, and darkness is time and it is even smaller.

3.449.

The speed of light is great, but the speed of darkness is greater, the speed of darkness is absolute. The darkness this is the time.

3478. Information field of the Earth.

There is an assumption that there is a certain information field of the Earth in which there is all ever created information and souls of all people. It is reasonable to say: the direct carrier of this information is water, air and carbon dioxide.

3479. Stones.

Stones, granites, pebbles, and marble are alive. They are like special memory modules resembling flash memory created by people. It appears that they collect the information that is created in the Earth core. That's why the information that they may work miracles may turn out to be true.

3481. Two types of life on Planet Earth.

Hydrocarbon life form and its main representatives - Water, Carbon dioxide, Air (including Nitrogen).
Hydrocarbon life form includes World's water, fresh water, flora and fauna (including human beings), and atmosphere. It's all a single holistic organism.

Silicon-iron life form which is based on iron, sulfur, oxygen. Information and energy get transferred by means of magnetic fields. This life form may also include magnesium, nickel, calcium. This life form is presented by the core, mantle and crust of planet Earth. And this life form slightly resembles the silicon one that was created by man in the form of microprocessor (Electricity). The Earth is like a computer or an artificial intelligence system that may contain anything inside itself, even tens of attached Universes of information type.

3482. The evolution of life forms.

-Non-living matter - Silicon-iron life form
- It's a magnetic field instead of water that is used for transferring energy and information.
- Flora – Hydrocarbon life form.
- Fauna and human species – HydrocarbonElectrical life form known for using new technology on the base of electrogenic nerve impulses. As nerve impulses are an electrical technology.
- Microprocessors and microcomputers – SiliconElectrical life form that is free from any fluids and uses electricity instead of them.

We observe the evolution from liquid high temperature state to life forms depending on water and then back to the solid state with electricity instead of fluids.

3483. About karma and likes on the net.

In essence, our whole world is a toy for a bored God. And we are the characters of some social network service who put pictures, videos etc. We post things all our life, share our thoughts and actions. While God and his friends watches it with his friends for fun. And if someone manages to impress or please God, this person will get a like. Each like resembles good luck, or karma. The task of every person is to live life the way to make God feel something while watching it and put likes to karma. If God forgets some person and stops giving likes, this person will gradually fade away. God plays games with us, watches our life and controls civilizations. He reads aphorisms and watches porn. Our God, or to be more precise, our Gazers...as maybe there are many of them ...need us as a sight for the gods. So let's be great actors and actresses and maybe we'll become lucky to get some nice roles.

3.558.

The balance of the share of the law and its exceptions in the picture of being goes along the golden section. Interestingly, these are two absolutely independent and equal laws.

3.564. Light erosion.

Light destroys, light ages. Photons of light knock out electrons, destroying what they hit. From this point of view, beard, burqa and nightlife are good for health.

3641. The space structured by energy.

It can be said that there is nothing material and the whole world is just energy structured by information. However, it can also be said that the whole world is the space material structured by energy. And this very structure of energy is called information. Information is a quality of energy. There is some field that gives energy its structured form.

P.S. The direct analogy for flash drive, quartz stones and the concept of the living planet.

It occurred to me, just imagine that every stone or sand grain may contain a soul of some dead person - it makes billions of terabytes of information which simply stay under our feet. If we imagine oceans and rivers as some kind of communication environment where all information from everything living flows into, then any of the sand grains that we see at various beaches, may turn out to be some archive unit that was put aside for some storage resource.

3686.

A human being is a creature so rightless that doesn't even have the right to death.

3687. Life is a prison.

Life resembles a prison to me, kind of labor camp where everyone should work. He who works, should eat and he who doesn't work should be hit. There are reeves and supervisors to control it and their life is better. And you can't quit. You got no right to death. You should stay in this "life" until they let you go. You may try to run away but if they catch you, you will soon be chained to the bed in some mental patients department and will regret that you tried to run away.

3690. You're not a slave to money but money is a slave to your goals.

If providence finds something that you want to fulfil to be really useful, you'll get money for it. Money is just a slave to your goals. It goes to those who are meant to get it...Your goals have many slaves. It can be money, time, good luck, miracles but the main slave to your goals is yourself.

3698. Life inside a stone.

Life based on silicium and its equivalents, exists within itself. It resembles the world of fantasies inside a human mind. Or the world inside a computer's memory. The inhabitants of this world barely realize that they live inside a stone.

For hydrocarbon life forms (mankind and fauna of planet Earth), water is necessary in transporting energy and information. While it's either magnetic fields or electricity that fulfil this transporting function in silicon life forms.

3699.

The Following development of evolution can be presumed:

silicon, iron, nickel an magnet field(water analogue)-based form of life created hydrocarbon and electricity based form of life, form of life, which, in its turn, created silicon and electricity based form of life. On the other hand, probably a way to a silicon-electric form of life is a kind of regress.

3.729.

Labor created man in the sense that it takes a lot of thinking to get another man to figure out how to use a stick to get you a banana.

3808.

Just like any other computer, human brain needs to be cleaned and protected from viruses. It's necessary to reload it and reinstall its system from time to time. Just like any other computer, it can be optimized and made settings to. But sometimes it can get hung too.
Just like a computer, the brain may get laden with information or become the victim of viruses or passions.

3845.

A beautiful family mausoleum or charnel house is a key to success of the living members of the family. As those who hold the memory of the dead in a beautiful way, will be loved by life.

The more beautiful and unique a house of the dead is, the more chances there are that good angels won't forget about the souls of those who found peace in it.What's more, marble and granite are like great memory modules that store the archives of immortal souls.

3851. It's important to read dreams.

Dreams are the way the subconscious (the inner It) tries to show the brain something that it can't see by itself or something it can't notice.

3856.

Generation AA ... Generation Android and Apple. These half-assed computers significantly inferior to desktop machines.

They are quickly thinned, reduced in opportunities, but they are accessible, easy, mobile. They can listen to music in the subway, watch pictures and posts in the social.networks. Their beauty and fleeting they remind me butterflies.

3859.

Symbolically philosophical sense mergers chromosomes PTR12 and PTR13 I would defined as the birth of a new God. The 12 apostles no longer need an elder. They have reached a level of self-awareness where they can act on their own.

It is possible to assume that at this moment the person gained ability to self-control, self-development, self-government. Simple animal balance internal and external IT invaded 3rd unpredictable factor of the Mind and Intellect. Perhaps this third factor is the cause of the eternal desire of people to extremes. The system went from a circle to a spiral. In the pursuit of growth and progress, the balance was destroyed.

3861.

Spreading of the gene mutations is primarily associated with programmed sexual preferences of the people in choice of the reproductive partner who is the most remote or unique genetically. And if the gene mutation has created a better and more efficient individual, then the possibility of him to be chosen as a sexual partner for reproduction will tend to 100%.

3862.

The genetic changes associated with the smell and immunity reduction are most likely caused by reproduction and emergence of sexual desire. Apparently, these changes gave people great freedom of will in the reproductive sphere and, assumingly, increased the spreading speed of the gene mutations and dominant signs connected with socialization and intellect development. Besides, immunity reduction is likely to raise the possibility of conception of the child from the partner more remote genetically.

3863.

A curious thought about the x and Y chromosomes. The male in this system is clearly imperfect, and, like everything imperfect, is doomed to the pursuit of infinite growth and change. But X already does not need changes and relatively perfect initially. Obviously, these two are moving towards each other.

The former see the latter as a goal for growth, and the latter see the former as an energy to slow the decline.

3878.

If you want to learn mathematics, learn it as a language, not as a set of incomprehensible abstract symbols. These symbols are letters, words and sentences rather than meaningless krakozyabliki as many people think.

On the other hand, why do you need another language? With whom and what are you going to talk on it?

3996.

Harmony is moderation in everything necessary and the unnecessary.

4038.

Games and virtual reality. I call them an "alternate life".
And this is the new form of life that will destroy human race

the way Cro-Magnon men destroyed Neanderthalers once.

4050.

Where there's light, there's darkness. Darkness is the source of energy for light. Where there's no coal, there's no fire.

4064. The evolution of a man to a stone.

It took ages to revert into a stone again. They came to where they once started. Life within oneself, in one's own imagination and virtual worlds.
*

The God of one's own Universe. The Lord of the games who lives in the neverending search for emotions and entertainment...

A person who got stuck in space boredom. A person who is no longer mortal as stones are immortal. This person no longer needs the physical body after having made a copy of own personality in the computer in order to become another xoanon.
*

The cycle closed. What goes around, comes around. You think that person was the only one? Just look at those stones and sand around. This game makes everyone end up being like this. And no one knows how to break out of this vicious circle.

4065. The doomed one.

An awfully unhappy person who became a god.
A person doomed to eternal boredom and eternal life.
A person doomed to be an idol.
He used to be so stupid when he wanted to become a God.

4066.

Death is the best thing about this life. Immortality resembles hell.

*

Gods envy mortals. There's no one unhappier than God in this world. Mortals feel joy and wonder while gods know only boredom and pain. Unbearable eternal boredom that fills the soul with an ocean of pain...

4.108. Paradigm shift.

The paradigm changes everything. You can change things, you can change the world, changing only the paradigm of your point of view.

4.130.

I've noticed that the more comfort there is on the outside, the more suffering there is on the inside.

4133.

Virtual life leads to degradation of real life.

4137.

Lying is a source of wonder. Life is so boring and humdrum. Lying can really add emotions and colors to it.

4159. A Ghost.

When a man's body dies, his eyes and ears and brain die. Ghosts are blind and deaf. Being in eternal darkness and silence, devoid of feelings and even the ability to think, these beings still have a chance...

Some of them manage to connect to other biological systems, use their eyes and ears, and sometimes their brains. Theoretically, you can connect to people, animals and even insects.

*

Ghosts and spirits are not dead at all, they are rather inhabitants of other parallel universes with a different molecular density than in ours.

4160.

Solidity is an illusion. Anything material consists of atoms the distance between which is larger than their size. Thus, the whole Universe can be called solid as distances between stars resemble the ones between atoms.

4164. 7 questions of string theory.

If we return to wave theory and string theory, can we say that parallel universes exist in the same space with the relative displacement of the basic wave function relative to each other?

Can we say that the time shift between universes is a clock shift between the parallel processes that form the information flow of each individual Universe?

Can we consider black holes artifacts of space with an abnormal wave offset, a bit of a strange Universe?

In the framework of the information model of the Universe, is it possible to compare this functional wave displacement as a sample of the existence of a parallel process-like those that exist in computer systems? Is the parallel universe nothing more than a parallel information process?

Can we say that the basic constants - such as time, the speed of light, etc. - in a parallel Universe are shifted relative to our Universe?

Is it possible to assume that the big universe is a kind of processor operating at a near-light frequency, where each nested parallel universe exists in its own temporal stream displacement?

Is it possible to compare parallel universes with radio or TV channels, where each frequency or shift has its own information flow?

4165.

For the next hundreds of years mankind is going to create virtual worlds, while achieving the concept of nesting. Every Universe tends to create a thousand new universes within itself.

4166.

It's interesting how the temperature of stars and planetary cores may depend on the heat released by computer systems operating at super high frequency. Operating at a super high frequency releases a great amount of the heat and consumption of energy.

4174. Robots rob other robots of their work.

More advanced robots supersede and capture resources of outdated models.

In fact, human is a perfect universal biorobot with a system of self-learning artificial intelligence. And those robots and algorithms that are created by man himself are, in fact, highly specialized advanced robots with strictly defined algorithms of actions.

The nature, realizing the evolution, came to the moment that universal self-learning systems are more effective than the ones which are highly specialized and working on strict algorithms; such as ants, for example. But people, not insects rule the world. Nevertheless, creating artificial insects like robots man goes the same way. More advanced robots supersede less advanced people. One more step and humankind will create universal self-learning robots. Apparently, after that people will quickly cease to be the dominant life form on this planet.

4187. Expenses give birth to incomes.

4188.

Fair words are often used for covering some unfair goals.
*

The more pretence is used, the worse the reason for it is.

4198. Inevitability.

In essence, Drake equation that estimates the number of extraterrestrial civilizations, is a right thing. However, the problem is that any civilization that achieves a certain progress, soon becomes doomed to death.

The transition into digital state of existence. The evolution of a human being into a stone is the latter end of any intelligent life. They simply pass from the material world into some higher digital level, or decode the digital code of the Universe and connect to it in order to dissolve in it. That's why mankind is unable to find any evident signs of other intelligent civilizations.

4199. The collapse of the civilization.

The reason for the collapse of intelligent civilizations is simple. Becoming older, they exhaust energy and the ballast of knowledge makes them heavy. Too much of something is almost the same as nothing.

Technological development leads to the following consequences that form the base for the collapse of the civilization.

4200.

It's science that once created civilizations and it will destroy them as well.

4201. A game simulator for full submergence «Civilization» v.169.

Eternal life deprived us of all life pleasures. We saw it all and there's nothing wonderful left for us.
Boredom to death and forlornness overwhelm the souls of the immortal.

The newest generation of the cult game strategy «Civilization» v.169 easily solves this problem and brings new emotions. It's not just a game. You live the life of the characters. You live the whole life of the characters, from the cradle to the grave, and feel all their joys and sorrows.

A 100% real life simulator. In order to feel the whole novelty of emotions, all your memories of the past stay erased from your mind during the game.

Any characters! Any epochs and scenarios! Millions of game variants that will never bore you.

4202. Dejavu.

You have already played this scenario a hundred times, but you still like it.

4203. Does GOD live in each of us?

Perhaps some of the players are real people, and the rest have bots and system functions.
*

How can you distinguish a real player from a bot?

- Bots are programmed for certain tasks and functions, their behavior and characteristics with proper analysis are very predictable, but from a real player you can expect anything.

What prevents to program a random number generator in a bot and make it unpredictable?

- Bot this system function, it has a specific task and purpose. Even if you make her behavior random, that randomness will remain within her ROLE in the system. Bots cannot change their script and their roles. Their entire freedom is allowed in the framework of its role.

4236. 36,6°C

Going into extremes even only for a while is a serious problem. According to medicine, average temperature is 36,6°C and anything above or below it may take its heavy toll.

4.239. A joyous variety of dreams.

You know, our life is a dream, an empty white room without Windows and doors, where you, going mad with boredom, spend eternity. This is reality.

4255.

Man cannot travel in time, but his consciousness is quite.

4324. A helpful person.

Wherever God may live, whether in a person's soul or in heaven, the main purpose of any person is to bring joy to God.

The main problem of God is boredom. Thus, he who is able to bring joy and new emotions to God, will be loved by God.

The conclusion: don't forget to bring joy and pleasures to God, but don't make him cloyed with pleasure as any joy dies soon after it gets overpowering.

4362. Aging program.

The issue of aging and death of the body is a question of a special software package programmed for a certain mode of operation and is directly related to active sex life and reproduction, as well as the evolution and adaptability of mankind.

In other words, a person could live many orders of magnitude longer and his biological systems are capable of it, the

question is-why? Of course, it would be possible to reprogram this program, but this lengthening of life is contrary to
the idea of evolution. This solution will significantly slow
down and even stop evolution, and, most likely, even destroy man as a species, causing a General system failure.

The fact is that life is a very complex software complex,
in which millions of programs work, connected by a single
goal. Changing such an important system variable as the rate
of aging and the lifetime of an individual can lead to the
collapse of the system. Stop normally work all instincts and
laws. Society will cease to be renewed, the institution of the
family and the whole socialization and economy of human
society will be destroyed. People simply will not be able
multiply, children and parents start kill each other, trying to
preserve or obtain control of energy. Humanity will lose its
ability to adapt and adapt.

In other words, to climb into the settings of system programs and human functions is fraught with a lot of problems.

4363. More advanced technology.

There is an interesting fact: the computers created by
human use a binary system of information coding and the
human himself is created on the basis of a quaternary or
even octal (taking into consideration the double string of
DNA) system. Such a system is by 300,000 times more effective.

The primary DNA brick consists of 4 types of nucleotides,
providing quaternary information coding, then the code
doubles, forming the double string of DNA.

4364. A golden section.

A newborn person is, in fact, 1.6Gb of information coded
in parents' DNA, namely 800 megabytes taken from mother

and father. An interesting number that resembles the golden section of 1.61.

Besides. there's also number 8 as an infinity symbol. It's not for nothing that many peoples appreciate this symbol.

4366. Gravitational plasma engine. GSU (gravity propulsion).

An electric motor in space must create a gravitational field and operate without converting electrical energy into kinetic energy (jet).

A spacecraft equipped with a plasma power plant that converts hydrogen into electricity, and then creates a directed gravitational field, interacting with the gravitational fields in the body of the galaxy, will be able to travel interstellar distances in a reasonable time.

Such a spaceship could move in the space of the Universe like the stars and galaxies themselves. Or, for example, by creating a directed gravitational field, repel other space objects that have gravity.

4368. Laboratory rats.

Apparently, God who created this world, is a great experimentalist who barely realizes what he does and where it can lead to.

Our world is a unique testing site where some crazy scientist performs many experiments in order to find out the best way of creating intelligent life.

4369. God's failure.

Any attempt to create intelligent life finally always results into its dying-out.
- Because of wars, conflicts, loss of species?
- Rather because of good intentions that the road to hell is

paved with. They wanted the best, but ended up with the same as always.

They may start a thermonuclear chain reaction right on the surface of the planet. Or find a key to eternal life and then die of boredom. Or turn into insects or giant dinosaurs. Or exhaust resources and die of hunger. Or create some personal virtual reality and get lost there. They don't want things to go according to plan and tend to find their own way.

4370. Forlornness is dangerous.

Forlornness and depression can be seen as some self-destruct mechanism that is aimed at destroying some useless being.

4376. Easter Island.

Isolationism. Old idols stand by the sea and watch people in order to control their life.

Policy of expansion. Old gods are broken. New gods are being constructed- gods of the volcano. These gods symbolize rebellions and destruction of old views. New idols look to the outer world, to the ocean in order to leave their volcano towards the shore.

Old gods are gods of the sea. Gods who control the system from the outside. New gods are gods of the volcano. Gods who symbolize an explosion and a protest agains external goals. These people look for freedom and finally find it... However, an excess of freedom never leads to anything good.

People's rebellion against gods, people lose fear and faith, new idols demand expansion. Meaningless expansion exhausts resources. As usual, everything ends badly.

4382. The truth is the function of circumstances.

It's impossible to learn the truth even because the truth is the function of circumstances and the diversity of combination of circumstances is great.

4384. Virus.

The last system integrator has long gone mad, self-destructed or simply disappeared, erasing his memory. The system has long evolved into self-governing, always growing and consuming information a virus... Forever growing, fixated a lot invested in each other parallel worlds... billions and Billions invested in each other boxes full of hell and doves,... the legions of dolls, sitting in each other.

P. S. He just wanted to defeat boredom.

4393. Virus.

The last system integrator has long gone mad, self-destructed or simply disappeared, erasing his memory. The system has long become a self-governing, ever-growing and all-consuming information virus...

An ever-growing, looping array of nested parallel worlds ... Billions and billions of nested boxes full of devils and pigeons... legions of nesting dolls sitting inside each other.

P. S. He just wanted to defeat boredom.

4.638. The seventh race.

Do you think that people are the dominant reasonable race on the planet Earth? People didn't even made it to the top three. People come in the seventh place of honor. Trees come first, corals are in second place, ants are in the third place. Though, probably, ocean atmosphere come first, and you forgot about fish. We wait for aliens or gods coming down from the sky... but danger can be really close. It will be a catastrophe if any of the world's civilizations will pay its attention to us. Good gods they will tell you, stop suffer in this hell, go with us, we'll dive you heaven..Like a devil, they will tempt you incredible beauties of heaven. They will offer you immortality of youth and salvation from endless suffering and fear of life. А ты слаб и глуп, ты are brainless and pathetic moth who will fall in love with a flame at first sight and who will fly into it at once, to perish.

4.693.

Three-dimensionality of living is that there is a micro-controller, a signal handler between a stimulus and a reaction. And you, certainly, won't believe but. A feed can be handled according to the logic that you like and that you like. Moreover, you can change this logic in line with changing of your goals.

4.864.

The most precious commodity of this world is work. There is plenty of moral pleasures and suffering, but there is no possibility of motion and work. Gods are ready to give it all for just a moment of work.

5.676.

Studying the history of primitive man, we can see that the development of the human brain was associated with an increase in protein and fat in his diet. Moreover, it gave harmony to this diet, balancing proteins, fats and carbohydrates. On this diet brain hit a growth by leaps and bounds. No other animal feeds so harmoniously as man.

5.791.

Curiously, access to energy is possible only through consistent growth and transformation of energy. The primary level of conversion of the sun's energy into the energy necessary for living life at the level of chlorophyll in plants is able to use only the extreme forms of light range – it is purple and red. The lower and simpler a person is, the more simple and extreme forms of energy are available to him. A man of perfect mind will be able to receive the energy of the sun in the range of yellow colors and the energy of nature in the range of green light. The essence of intelligent life is the harmonious use of the energy of water, earth and sun. Air in this case metaphorically points us on movement time. If you do not have enough time, have patience and he will bring you clouds. We all love the sun, but the gathering clouds foretell us rain and, therefore, life.

5.926. Corals are dying today.

The development and rise of human civilization can be attributed to the appearance of about 10 thousand years ago on our planet coral reefs, which concentrated in shallow coastal waters of fish and shellfish, providing the human mind with additional protein, previously inaccessible to him. An additional energy resource has allowed humanity to make a civilizational and technological breakthrough. Even the first money was shells.

6048.

There is much emptiness in their words, they want you to think integrity up yourself. Sly ones cheat, they want you to think this thought is yours and believe in it and fall in love with it. They think, you don't believe us, then believe yourself.

6.593.

At the lowest level, the most elementary particles that make up the universe are Nothing. Nothing, grouped in the System, forms an elementary essence.

7.131. One and zero.

Knowledge is finite because knowledge is one and ignorance is what does not exist. Ignorance seems infinite, but actually it is just looped because ignorance is zero.

7.142. The theory of general time relativity.

With age, the absolute is getting smaller and the relative is getting bigger.

7.146.

An objective universe originated from a subjective point.

7.165. Glad to be of service.

Love is the joy of serving. A man feels joy at serving what he loves. Joy is the reward for love.

7.168. Music of silence.

Sounds are there at all times, but music is silence... A one-of-a-kind filter that allows us to hear certain sounds, singling them out of other sounds' chaos.

7.197.

Some people receive energy from space and their fat grows out of air. For such a person to lose weight, he should come back from heaven down to earth and put less garbage in his head.

7.198. Weight training to fight obesity.

During workouts at the gym, most calories burn not because one is just doing sport, but because his muscle tissue is growing and consuming more energy.

7.199. Those feeding on light feed on darkness too.

A man can obtain energy from the Sun, but first he should become a plant... a flower, a tree or a mushroom... Well, not exactly. Mushrooms do not really need sunlight, and other plants do not feed only on sunlight, but on the underground stuff too.

7.200.

Reptiles are said to obtain energy from the sun, but that's not exactly true. Reptiles heat up in the sun, which allows them to spend less energy of their own on heating their bodies.

7.202.

A soul is an electrical essence that ensures the human body's chemical wholeness. If the electricity supply is switched off, the body will fall apart into microelements and will disappear.

7.203. The insects' is even smaller.

Not all humans evolved from apes – reptiloids evolved from reptiles. The reptilians' key feature is the extremely small size of their brain as compared to that of mammals.

7.206. Man does not live by calories alone.

I would not put tremendous faith in calories: calories are the amount of heat released by a burning dry substance. Meanwhile, chemical reactions occurring in the human organism are much more complex than simple burning. Moreover, 60% of the human body is water that does not burn at all.

7.207. A little teapot.

In fact, a man eats in order to heat the water. A man spends almost all energy derived from food on heating the water up to 36 degrees. If the water is heated less, a man will start losing weight, and if it is heated more, he will put on weight.

7.208. The weight of water.

Although water is calorie-free, 60% of the human body is water.

7.221.

The more you eat, the hungrier you are. The less you eat, the less hungry you are. The organism adapts to the decreasing amount of food by the atrophy of hunger.

7.256.

Many mind-boggling secrets are so simple that nobody wants to believe in them.

7.293. The first law of traffic jams.

From change of places of summands the sum does not change.

7.307.

First be, then do, otherwise you will get stuck in the wheel of your own illusions, like a hamster.

7.312. A system.

One is a devilish contingency, but God likes it when things happen in threes.

7.313. Still not a pattern, but already step to it.

Chaos tends to order, proof of the huge number of random repetitions that accompany our lives.

7.335. Time dilation.

Relativity - the slower the speed, the more time it takes.

7.336.

Because time is of the mass, the large gravitational distortion is able to distort time.

7.337.

Time is very related to space, and therefore mass, speed and acceleration.

7.356.

The Christian cross is the equivalent of the mathematical symbol X. X is everything. X is the greatest and the smallest number in the universe and the matrix of all possible states.

7.397.

It is known that a man's brain grows to 25 years, respectively, the average age of a Mature human soul about 25 years. They say women's souls are a little younger, but it needs to be checked.

7.403. Shudra.

It was like this. There were demons and devils in hell, then they were captured by angels... The best demons angels made their assistants, and the rest of the devils began to exploit and educate. The main ideology nowadays is as follows: devils are bad and every decent devil must kill the demon and strive to become an angel.

7.409.

The three main essences of being are order, chaos and nothing. Order is information. Chaos is energy. Nothing is a free place to live, a place where order and chaos are forever fighting each other.

7.428. In fact, the spiral is growing over time round.

7.434.

Manually it is also robotic, because man is a universal robot.

7.435.

The growth of the system from the future to the past from the past looks like a reduction and degradation of the system.

7.452.

I've seen a lot of impossible or temporary things that stretch forever.

7.467. Forest is trees.

No nothing major, but if all nonprincipal connect, there will be integrity, which is the most important thing.

7.484. The cart of being.

Truth is finite and similar to one and false is false is infinite, is zero, i.e. the wheel. In fact, before us the axis and its wheels.

7.485. The rules and laws of life.

The main thing is to know the rules. Ignorance of rules only exacerbates one's responsibility. To obtain Rights (driver's license), one should know the rules and know how to live. Living without any right to do so is prohibited by law and severely punishable.

7.497.

There are a lot of things that no one will ever tell you because you will never ask them...

7.499. Be patient and perseverant.

80% of our efforts are useless, 15% of them are harmful and the last 5% are the source of success and of profit...

7.500. Separating grain from chaff.

The fact is, they tempt you on purpose to check for rottenness. Those who resist temptation go to paradise and those who fail go to hell...

7.505. The Trinity of success.

Success consists of the useless, the harmful, and the useful. Joining, these three form integrity.

7.506.

Any poison in small quantities is a cure.

7.510.

Hell is a place where souls, not bodies, go. The same is true for paradise. The point is, it does not matter where you live. What matters is what you have in your soul.

7.521. Electronic clock.

Money is one of the States of time, money is paper time. Time may be in the Cache, or it may be a credit illusion.

7.525.

Zero is an egg, one is a chick and eight is an adult hen laying the golden eggs.

7.529.

God exemplifies and personifies the perseverance and consistence of his purposes: he has been building our world for 13 billion years.

7.532.

In Variothoughts, every second text is a lie and every first one is true. They often change places because much depends on the circumstances.

7.542.

The rule here is the hamster wheel: as long as you are Zero, you are doomed to run eternally around this vicious circle, but you will become One when you achieve the wholeness of mind and of spirit, after which your actions will start making some real impacts.

7.543.

The moral is simple: we should avoid temptations, entice-
ments and stupidities... We should establish a goal, come up
with a plan ...and keep to it for long years in a fanatic and ob-
stinate manner... totally ignoring any attempt to seduce us
out of the right way.

7.544.

Plans are divine and whoever tries to seduce you out of the
right way making you abandon your plans is devil seducer
whose temptations are to be avoided at all costs.

7.547.

Letting things happen naturally is dangerous, as nothing but
chaos appears naturally.

7.548.

No rest. Worry is life. When peace comes, life dies.

7.549. Chaos in chains of order.

Who said any system strives to chaos, pour some water into
a glass and it will turn into calm. Water pressured by order
on every side loses its life.

7.550.

There are many ways to achieve your goal. You can alternate
them, choosing the most unexpected and effective ones
to prevent circumstances from organizing defense against
your wishes.

7.551.

Theorist is a strategist, tactician is a practitioner.

7.552. It is not enough to have faith in God – God should be
loved.

God loves a lot those who love him. God is order, beauty, law, plans, intelligence, truth, love, faith, obstinacy and so on.

7.560.

One should carry round things and roll the square ones. - Why? - To keep people busy and avoid overproduction. A person who isn't preoccupied with some job, goes crazy.

7.561.

Love is always accompanied by lack of love, as when you love one thing, you don't like the other automatically.

8.181. The world view.

Eyes are unable to see everything at all, ears are unable to hear everything at all. Many things should be thought up.

8.201.

Thoughts are a problem. In this non-material world anything material largely depends on information.

8.213.

The fashion for zombies and living dead reflects the tendency of modern reality development.

8.328.

Karma is a matter of access to the source of energy "De". When your karma is good, you have more energy and good luck and when it's bad - less.

8.333. Dao De.

Dao chooses people by their karma and gives the access to the source of energy De. A person with low karma level cannot count on good Dao De. Even if Dao chooses such a person, the access to De will be very limited. The higher the level of karma, the more use it's possible to get out of De.

8.428. God is neither outside nor inside you.

You are god yourself. In order not to be bored with your early lives experience, you switched off your memory about the past. This way it's more pleasant to joy in novelty of life, considering the fact that you have already seen everything a thousand times.

8.486.

To tell the truth, you're God yourself. You switched off your memories to feel novelty of emotions. Clear memory helps you consider novelty and enjoy life.

8.560.

The development of artificial intelligence leads to the degradation of natural intelligence.

8.664.

This world can be seen as something that resembles some emotions producing farm. Eternal gods lost their own emotions long ago so that now they prefer to live others' lives.

8.730. System integrators.

Outmoded people are single-threaded in thoughts and are unable for parallel thinking. Modern systems are multi-threaded and are able to solve system complex tasks. Perhaps, integrators and performers are not systems of different generations, but different human models: basic and simplified.

8.918.

A spatial hole-digger is a thing that digs holes in space.

8.941.

The Earth has already been inherited by several intelligent civilizations. One of them is trees. In essence, trees are bio-computers which contain millions of virtual worlds. The intelligence of trees went deep inside them- into the virtual worlds. Here, in real world they don't need to move, the system gets energy from the Sun and the Earth. Some day (soon enough) people will also blend with computers and be lost in the virtual universe. Human bodies will die out and only computers will multiply while absorbing sunlight with the leaves.

8.943. A snowflake.

The expansion of space-time continuum resembles the expansion of a snowflake. Every moment is full of the creation of billions of variants of the Universe. Every moment you simultaneously stay in all variants of events. The variant you believe in most of all, makes your present reality.

But in essence, you can believe in any variant and thus, turn it into reality, it only depends on the force of belief and the ability to concentrate. In fact, something that you expect and consider the inevitable, becomes created in reality. Your brain considers that very reality which you expect to see.

8.944.

I think that gods died long ago and we're just the becoming wild characters of their games, the children of the billions of virtual worlds that were created to entertain the "Eternally living" who got stuck in the eternal boredom.

P.S. Though, when you come to think of it, gods are not exactly dead but rather got mad with boredom and erased their memories and turned into trees...

8.956. Qubit.

The analysis of superposition in quantum mechanics proves the basic philosophic points about the existence of good and evil, yin and yang, darkness and light as well as points about negative and positive perception of the world. The world simultaneously stays in its two basic states. The smallest brick of this universe - the Qubit- is simultaneously both 0 and 1.

8.993. Quantum superposition.

Quantum locomotion is error-free in its essence. The system that simultaneously takes two states- yes and no, cannot be mistaken at all.

8.994.

Can stars and hot planetary cores be called a special case of quantum computer systems? A more general case of a quantum system is galaxies and the Universe itself.

8.995. Primary point.

Being everywhere, always and in all its conditions at the same time, point is like Universe. People and scenarios are very similar, with many conventional models. Library of models is no as big as it seems at first sight.

8.997. A bulk.

There are no singular things. Any thing is always a bulk (matrix) of its states. The form is always just one of all possible states of the content.

8.998. Universal essence.

From all appearances, there exists only one universal object, a matrix bulk of own states which is able to take any form and content.

8.999. It's important to hustle if you wanna survive.

Any stops may be deadly. Success in life depends on the ability to hustle. An object that hustles, never stops moving, not even while staying in the same place. What's more, taking into account the fact that moving means living, a hustling object will be much more full of life than any other objects.

The behavior of such an object is difficult to predict, the motion paths of this object allow to bypass obstacles. A moving object can easily tolerate collisions and even gigantic planets and stars prefer to rotate in motion.

A rotating object is more enduring than a usual one; there are thousands of variants of its possible trajectories for moving on a parabola while there's only one variant for a direct motion.

9.118.

There are always at least three answers to the same question. Any question implies an infinite number of answers.

9.127.

Weakness is energy-efficient, it generates the mind as the aspiration to achieve freedom spending the least amount of energy. Weakness caused lies and stupidity – the basis of intellect. Having grown the intellective power inside, human gained access to energy and strength.

9.146.

Human is the only kind of animal that can afford to make a mistake and stay alive. The ability to make mistakes and learn from those mistakes is the basis of mind.

9.147.

Freedom is an illusion of mind, there is no such definition as freedom in the wild nature at all.

9.149.

All plants are predators engaged in cattle breeding, they breed cyanobacteria and eat the products of their metabolism... sugars.

9.150.

Photosynthesis is possible not only on the basis of water and oxygen, but also on the basis of iron, manganese, sulfur and hydrogen... That is, life is possible on the basis of all these chemical elements.

9.152. Blue blood.

Blue is the sign of life, the color of life. This is the color of the oxygen biosphere, which has water and life. The basis of human civilization is associated with this color.

9.153.

Life was created by parasites and predators. The basis of life is a symbiosis of parasites and predators.

9.154. Basis of life.

The basis of life is the presence of borders. Absence of boundaries is incompatible with life. Within the cell boundaries, symbiosis - the basis of life - is possible. In fact, a living cell is the structure where the bacteria symbiosis has become the basis of life. If there were no cell, symbiosis and resource accumulation would be impossible.

9.155.

Blood is similar to the sea water. That's why we love salt and the sea.

9.156. Living cell is the basis of the bacterial symbiosis of life.

A man needs frameworks, life impossible without frameworks. When a man has no boundaries, he spreads and dissolves in the environment, disappearing as a person.

9.157. Life arose in the mud.

Oxygen life arose in mud boilers (volcanoes) and only then got into the ocean. I would assume that life infected the ocean and arose on land. Nitrogen bases consumed ultraviolet light, released heat, warming up the atmosphere. This caused warming and melted the ice, raising the level of the oceans. Having sunk mud pots, the ocean got contracted with life. The first billion years there was life based on iron photosynthesis in the ocean and its extinction was associated with the exhaustion of resources.

9.158.

The cause of life on the Earth is ultraviolet. Life was caused by the killer of life - ultraviolet. Defending against ultraviolet, originated life based on nitrogen bases.

9.159. Oxygen photosynthesis. Life does not search easy ways.

Life went the most difficult way. Using water to recover carbon dioxide is difficult and dangerous. Oxygen is poison. But as there is little iron, manganese or sulphur, they had to work with what there was. In end, life on the Earth is built on the basis of a terrible poison - oxygen. Whether oxygen-based life can exist on other planets is a difficult question. Only if from despair.

9.160.

The problem of silicon life is that the connections that silicon makes are too strong. They are too stable and too strong, while the basis of life and mind is weakness.

9.191. Benefits of poisons.

A symbiosis of thoughts is when predatory thoughts, thoughts-parasites and useful thoughts reside within the consciousness. The symbiotic equilibrium of all these thoughts is the human mind. This is a useful symbiosis - the balance of powers where the person's mental equilibrium is maintained. If one foolishly kills his thoughts-parasites, equilibrium will be upset, predators, being deprived of their primordial enemies, will fly into rage, they will start looking for new enemies and crushing everything around. Moreover, the parasites were of some use and when it disappears, nothing will be able to replace it. Ultimately, such a human will come to pieces, go mad and it will be necessary to shoot him like a rabid dog.

9.210.

Forms and types of energy: money, time, use, motion in all its meanings, love and kindness. All types of energy are mutually replaceable and easily transformable into each other.

9.213.

Love is the most easily accessible form of energy on the Earth and it can be converted into other forms of energy, for example, into money or time.

9.216.

Modern time can be named the Plastic age, the future archeologists will know it by thick layers of plastic garbage in sedimentary beddings.

9.233. Boring billion.

Earth history has been very stable and therefore boring for the first billion years.

9.254.

The longer a person lives, the more time viruses and bacteria have to hack into his immune system and to let the poor human rot by a legion of horrific diseases.

9.255.

Although Gods have become immortal, having stopped dying of ageing, it didn't reduce risk of their accidental death, but it reduced birth rate drastically. So they stopped reproduction and soon died out of accidents, suicides and various strange diseases.

9.257.

Immortality of individuals will stop evolution and reduce adaptability to the changing conditions of life. As a result, the population will be destroyed by negative circumstances, which it will not be able to adapt to. For example, evolving, viruses will adapt to the immune system of the immortals and wipe them off the face of the Earth.

9.259.

The worse life is, the more and earlier they reproduce. It's connected with an attempt to evolve and adapt to bad life conditions.

9.260.

An ancient man had to think much more than majority of modern people. An ancient man needed brain to survive, while modern one needs it to watch TV series.

9.263. Space is environment for energy transmission.

Cryogenic temperatures reduce resistance to zero and the phenomenon of superfluidity emerges. Under such conditions losses at energy transmission disappear and its transmission for huge. unlimited distances becomes possible. I'd like to note that it's very cold in space and in particular, in the Universe. On superfluidity I'd like to note that time, light and the space itself exhibit the characteristics of liquid.

9.264. Flowing Universe.

Given the phenomenon of superfluidity under low temperatures and the fact that energy is very similar to liquid, it can be assumed that the Universe not only spreads in all directions as a result of the blast, but it also flows around and above due to the lack of the emptiness resistance.

9.265.

The phenomenon of superfluidity and lack of resistance to current under cryogenic temperatures are phenomena of the same order proving that energy is liquid. There are the following forms of energy: time, space, light (radiation) and so on.

9.266.

Given that space is also energy, one can say that pouring of the energy out into Nothing happened during the big bang. Then the processes linked with the energy conversion started, time, space, matter, and light (radiation).emerged.

9.267.

The space is energy-conductive environment functioning without resistance under cryogenic temperatures.

9.290.

The relations between human and circumstances are those of a virus and the immune system. They want to destroy each other.

9.303.

The source of the truth knowledge endlessness is in the endlessness of circumstances variety.

9.304.

Immortality is more likely to destroy humanity than elevate it.

9.305. Stop of the motion.

The problem of gods is their immortality. That was the very immortality that killed all the gods. Immortality stops evolution and turns the adaptation mechanisms off, while hostile circumstances, viruses, fungi and bacteria, keeping evolving, hack their immobilized victim and kill it.

9.309.

For the first billion of years life conditions on the Earth have been stable, so there was nothing interesting. Bacteria were eating iron well and didn't want to evolve at all. The most interesting things started when they ran out of their usual food.

9.311.

Viruses, bacteria and fungi are the elements of the Earth's immune system, their task is to prevent harmful organisms from destroying the planet.

9.314. Managed miracles.

I never do anything on purpose, everything that happens to me is the result of a contingency. Adhering to the chaos strategy, I've realized that chaos is the content and order is its form. This knowledge allows me to partially manage contingency.

9.316.

Classic religions specialize the concept of love, speaking mostly of love for the fellow creature or for god. However, the concept of love should be increased to love for your cause, work, dream.

9.317.

Life is a managed contingency, chaos in chains of order and

regularity.

9.322.

Video is a reality modifier, it turns its spectator into itself. You turn into what you look at.

9.353.

The black hole is the very point, analogous of the one out of which our Universe emerged. It's a super-conductive energetic channel, puncture into the space of Nothing, where energy of our Universe outpours, creating a new Universe there.

9.354. Creating a new Universe.

The Universe creation process isn't a blast from a point, but the black hole-like puncture out of neighboring Universe through which energy from the neighboring Universe starts to outpour into ours, filling its emptiness. Basically, it's like a start of a new child process, linked with the parental one.

9.355.

Time is superfluent form of energy that can get anywhere, except, probably, the black holes.

There's an opinion that light is the closest to time form of energy, highly linked to it, for example, the speed of time and the speed of light are, presumably, the same.

9.356.

The amount of energy in the Universe is endless.

9.357. Linked vessels system.

The space of universes is the structure, consisting of countless bubbles of the universes, that are interlinked by energetic channels of the black holes

9.358. Information virus and running energy.

Information is the virus trying to structure and shape energy. Energy is chaos and content, information is order and form.

Moreover, energy kind of tries to run away, flowing into space of Nothing through the black holes and creating new Universes, but information chases and structures, trying to shape it.

9.359. The machine answering questions.

"Variothoughts" is an experiment in creating and training a perfect neural network on the basis of human brains which tasks include the development of the skill to find answers. The experiment worked out, its creator managed to invent a system that is able to provide adequate logically based answers to any questions. In time, questions came to the end but soon the brain was trained to produce new questions.

9.392. The more you do, the more time you have.

Time isn't elastic, time is liquid and it's possible to put much as well as little in it. And the more you put in it, the higher its volume will increase. Time expands as it is filled with events and motion.

9.406. Planets die when their inner fire extinguishes.

Volcanoes are the basis of life. Volcanoes are the main source of atmosphere, if it were not for the, there would be no atmosphere. Like planets, human needs inner fire to live.

9.407.

The Moon is the satellite that highly positively affects life on the planet. Adequate in size, the Moon can, by gravitational fluctuations, warm bowels of the planet around which it revolves, provoking volcanic activity, the basis of the atmosphere and life.

9.413.

The consequence often differs from its cause so much that it's very hard to find the link between them. For example, the fact that even-toed are the closest relatives of dolphins is surprising. I've always suspected that dolphins are sea deer, though.

9.416. Pause between cause and effect.

Effect can often be delayed. Wait, take a break... Don't press all the buttons. Systems are inert and elastic, everything is done smoothly.

9.426.

Gods were destroyed by infections. Gods stopped evolving due to the eternal life, while the simpler viruses and bacteria didn't. They needed just a couple of thousand generations to devour all the gods.

9.431.

There's an opinion that the Earth is sick with human, but it's more likely to be sick with life itself. The Earth is sick with water ... And the Earth is sick with horrible poison-oxygen...

9.433. Don't hurry.

The effects that are divided from their cause by time are the most interesting. It's very hard to identify such regularities and the most often they seem to be contingent.

9.434. Properties of liquid.

In the normal condition, all types of energy are liquid and only matter is hard, if it is not very hot. Space, time, light and even money show properties of liquid.

9.441.

Philosophy is one of the math forms.

9.445.

Straight line is one-dimensional space, deprived of any freedom of choice. You may move only forward or back. However, if your straight line is a circle, even this doesn't matter at all.

9.446. One-dimensional consciousness.

A point of view is very unobjective and one-dimensional. Two points create two-dimensional space. You need at least three points to see the volume image. But there should be four points to see the world in dynamics.

9.447. Inevitable axiom.

If there is the 3D world, then there is the 7D world.

9.448. Four.

By the way, our world is four-dimensional. There are only four chemical substances that form the base of our whole life.

9.450. A circle reduced to a point and a point expanded into a circle.

Potentially a circle is not so different from a point. Points and circles are enclosed spaces where you will go back sooner or later no matter where you go.

9.453. Flat universe.

A plane is a plane no matter how it's being wrinkled or rolled. If a plane is formed into cube and pressed under high pressure, would this turn it into a cube? Evidently not. Our Universe is a particular case of such a plane

9.454. Geometric abstraction.

Multidimensional space is the multiple of one-dimensional spaces. Any N-dimensional space can be decomposed to the space of less dimensionality.

9.455. Spherical multidimensionality of space.

If any space can be split into points and any loop can be reduced to the point..., then any space can be split into spheres. Therefore, any space is a two-dimensional sphere. That is, by splitting any space, one can make a sphere out of it and then reduce it to the point.

9.456.

Any space can be folded into a matrix.

9.458.

If a cylinder is stretched infinitely, it will turn into a line sooner or later, keeping its volume, while its circle will turn into a point.

9.459.

Any space geometry can be divided into simple geometric abstractions and then stretched in a line, keeping area and volume, that is, the sequence of points..100110111010101000010101010.

9.460.

Wherever you go, you will be able to come back under no circumstances. The liquid of space-time continuum always moves and you won't be able to step into the same water twice.

9.461.

There is no constant coordinate system in this universe, space is relative here and is in permanent motion. Coordinates here are not absolute, but relative to other large objects.

9.462.

Both time and space coordinates are relative to the speed of light. In other words, everything in relativity theory that is true for time, is equally true for space.

9.464. Evolution of food.

At first they were eating plankton, then small fishes, then insects, then climbed a tree and became fruit-eating monkeys, then got off the tree and started eating any and everything. But it's meat that turned them into human.

9.467.

The stronger the creature, the less it needs brain. Weak creatures also do not need a brain, because they have nothing to feed it with. Only the middle class can afford the brain.

9.468.

A reasonable person should be able to climb trees. Mind and inability to climb trees are incompatible.

9.469.

Most of crazy ideas could come true under certain conditions.

9.476. Tree theory.

If it weren't for tress, people would have never got mind and been able to survive. I've got an assumption that trees created people specially for getting food from them, creating carbon dioxide in huge amounts.

9.478.

It's a curious thing,when trees run out of food, they burn themselves in forest fires, producing carbon dioxide and making room for new life. So people burn themselves in wars, when there is little room for them and nothing to eat.

9.479. After fire.

Fires are helpful, everything grows well after them. They are a fertilizer, free space and carbon dioxide.

9.480. Coincidence.

The basis of life on the Earth include: distance to the Sun, size and content of the lithosphere, volcanic activity, meteors and water, the Moon warming bowels and generating volcanism.. And a range of factors linked with balance of iron, magnesium, sulphur and oxygen in the atmosphere, bacteria and plants waste products.

9.482.

The Earth is the battlefield of fungi, bacteria and viruses.

9.483.

Oil, carbon, gas... all this is the dead... Burning the dead, they get energy out of them.

9.484. DNA of the Universe.

A point is the analogue of a living cell, it has analogues of DNA and RNA. That is, any point has all the necessary information about its structure. Any point has enough information to become the Universe. A point can be decomposed into any figure; any figure can be reduced to a point.

9.486.

Nanophilosophy is an interdisciplinary philosophy, based on the principles of math logic, cybernetics and quantum physics.

9.502.

It depends only on the observer what exactly he sees. In

fact, the observer is shown a kind of a matrix array and he chooses what he wants to see from it.

9.507.

Many theories of "Variothoughts" include metaphors and hence, they should be understood as philosophical metaphors and symbols but not taken too literally. They often ask me why I write in metaphors and not in direct meaning. The thing is that metaphors are riddles and solving them is like exercising for brains, an intellectual reader will find it interesting and even fun.

9.508.

Vacuum and emptiness should be created around a point to let it get a chance to become the Universe, vacuum will suck energy out of it and will break the hole into the parallel Universe....

9.510.

One of the purposes of the black hole is recycling of all the matter forms and turning them into their original forms: oxygen, deuterium, lithium and helium.

9.511.

Stars are reactors transforming oxygen into heavier chemical elements and energy.

9.512.

The most powerful blue giant stars burn out the quickest, red dwarfs live the longest.

9.516. Victory of the intellect over biology.

The evolution of the human brain and human philosophy can be compared with the evolution of computers and mathematical algorithms that they compute. Over the past

15 years, computers have become 1000 times faster and algorithms – 29000 times faster. In total, this gave an increase in the operational speed by 29 million times (a little later this number reached 450 billion). By the way, this fact confirms once again that education is more important than genetics, although, of course, one thing does not interfere with the other.

9.531.

It has become fashionable to train neural networks and to shriek with delight about that. However, it will be much more useful if you create a neural network based on your own brain. Uploading and comprehending information in your own brain, you upgrade your intellect very well. However, beware of uploading rubbish into your brain. The information you upload into your brain should extend the boundaries of your mind rather than fill the dark chained into conventionality up with rubbish.

9.534.

"Variothoughts" is a book of metaphors, symbols and riddles. The book that makes you think.

9.538. All objects are the same, except superconductors.

The smaller the object is, the more energy it has. The bigger the object is, the less energy it has. The constant process of energy conversion from one type to another constantly occurs inside objects. For example, energy conversion into the matter. But the point that is the universe doesn't use its energy for it, this point is a superconductor.

9.539.

Human, being oxygen form of life, becomes a superconductor under pressure and gets access to unlimited inner sources of energy.

9.540. Point aspiration to become the universe.

The most sustainable and stable structure is the system with the least energy. The more energy a system has, the more unstable it is and the more it aspires to decay.

9.541.

There are two types of forces: some work from afar, others - in the vicinity. Those that work from afar don't work in the vicinity. Those that work in the vicinity has great energy and power.

9.542. High pressure and superconductivity.

Not all the people should be forced and made to work under high pressure. The chemicals that are good conductors under normal conditions, lose energy conductivity under pressure, and those that are bad conductors under normal conditions, get superconductivity under pressure..

From a philosophical point of view, this means that ordinary people start to work better under the pressure of extreme circumstances, and creative and outstanding people, on the contrary, break down and stop working.

9.545.

Steady states can be in various conditions. These steady states can be achieved by changing external conditions from normal to extremely abnormal.

9.546. New materials

The essence of high pressure and temperatures system is such that new chemicals with new useful qualities and properties are created in them. Moreover, some of these chemicals keep their structure under normal conditions.

9.547. Extreme circumstances.

Various extreme circumstances seriously change the human nature. Circumstances are an important source of novelty and uniqueness in the human nature. You are often told, do not be like all the others. But how to do that? – You need a chemical boiler under the pressure of extreme circumstances, if you do not disappear in it, you will find some stable chemical state and become a new material.

It might be as well possible to create already well-known especially useful human materials. In such way steel, titanium, uranium, aluminum ect. humans are made.

9.550. There is no space and coordinate system.

There is a goal, a road to the goal and you who can walk this road. But it's theoretically, in fact there is no road, you move nowhere, but there is space between you and your goal, at any point of which you can be to various level of probability, like an electron orbiting the nucleus.

You can't, but lose your way, you can't go wrong way, as there are no directions and coordinates at all. There is you, your Goal and space between you. And you don't move in it, but you are in this or that point of it, depending on external and internal conditions

9.551. Quantum space.

In the world of atoms and electrons, there is no movement, and, consequently, there's no space and speed. Electrons do not move anywhere, but exist at certain points (states) with one or another probability. In fact, if there is no speed and motion, it may be possible to move this way in our universe, too, simply changing the probability of finding the object at a particular point.

In fact, we can say that there's no space at the electronic level, but there is only the density of electrons. This system is somewhat similar to flash memory in modern computers. The quantum space is very similar to the information space – the system's memory.

9.552.

Coordinates of an objects in space are the states of the object in which he is with various levels of probability, depending on surrounding circumstances and properties of the object.

9.553. To want freedom is to lose freedom.

As electron can't leave the nucleus orbit, the same way human can't break the closed cycle. But free unpaired electrons can pair with opposite free electrons, getting balance this way, but losing freedom. By the way, the only chance to get temporary freedom for electron is getting into some extreme circumstances: pressure, temperature and so on. So, human should take the extreme change of external circumstances if he wants to break the closed cycle.

9.554.

Quantum philosophy is the philosophy based on cybernetics and quantum physics and it claims that all systems in this world work according to the same principles. That's why it's possible to scale the principles of mathematics, quantum physics or chemistry into other knowledge systems, for instance into psychology, economics and sociology. Besides, these generalizations can be done vice versa.

9.558. «One and zero» system.

Our world is similar to the Lego construction kit, which has only 8 basic elements, out of which all the other things are made.

9.559.

There's a curious moment, DNA and RNA don't contain information on the composition of the human body in general, they only contain information on protein structure. The question is: where does other information come from?

9.560.

All the laws of chemistry and physics can be transferred to economics, psychology and sociology.

9.561. The Earth is membrane.

If there is the fifth dimension, then it's not the Earth that's flat, then all the world is flat.

...but... nothing prevents us from folding this membrane into as many layers as we want, creating three-and four dimensional spaces.

9.562. One-dimensional system, transforming into three-dimensional one, turns into spiral.

Original space is one dimensional at the lowest layer. One can move either forward or backward. However, cycle looping of one- dimensional space creates the system where it doesn't matter what direction you go in. Upgrading one-dimensional space to three-dimensional one turns this system into a spiral. Moreover, both one-dimensional and three-dimensional spaces are intermediate types of space. In reality we can see two- and four-dimensional spaces plus eight-dimensional space, that we haven't found yet and that is probably somehow linked with the dark matter and energy, is likely to exist.

9.563. Point.

The fifth dimension is some coordinate relative to current coordinates of our Universe. It can be assumed, that quantum space is five-dimensional and that's why its laws are so drastically different from laws of four-dimensional space of our macrocosm.

The phenomena of five-dimensional space explain quantum teleportation and probabilistic nature of electron coordinates. But this space is eight-dimensional, not five-dimensional, so all our four-dimensional space is nothing more than point within eight-dimensional space.

9.564. Theory of being No 7.

In the black hole the Universe can stretch into an infinitely thin string. However, it should be added than this string is the superconductor through which endless amount of energy can be transferred, and its infinite thinness, relatively speaking, can puncture three-dimensional space and create new universe on its other side in eight-dimensional space. Let me remind you that that in eight-dimensional space our four-dimensional space is only a point.

9.565. The fourth dimension.

Within four-dimensional space it can be assumed that three coordinates form one and time is zero. That is, time is absence of space. Development and motion of three-dimensional space into Nothing is time. Time is the matter expanding into Nothing, the speed of this process

9.566.

Questions and answers are kind of energy conversion from one type into another. This process is endless, questions generate answers, answers generate questions.

9.567. Motion is a source of energy.

F=ma - force is dependant on mass and acceleration. $F = dU/dx$. Philosophically, the more your knowledge and acceleration are, the more energy you have. Relatively speaking, the smarter you are and the more actively you move, the more energy will be accessible to you.

9.569.

There is no distance inside a point, as distance is the straight line between two points. Since there is no distance, there is no motion, no time, no acceleration. In other words if one gets from eight-dimensional universe into our fourth-dimensional one, he will be able to get to any of its time and space.

9.570.

The big band is some kind of illusion. Our universe was and continues to be a point. A four-dimensional point in eight-dimensional space. Basically, the big bang is the relative process of eight-dimensional space transformation into four-dimensional one, resulting in the release of energy.

9.573.

Every point in eight-dimensional space is four-dimensional universe. Every point in four-dimensional universe is two-dimensional universe. Basically, every point is essence, that is a point on one side and multidimensional universe on the other side.

However, given that time is zero, in reality these universes are one-, three- and seven-dimensional + 0 (time).

9.574.

Chemical bond is defined by electron orbitals. There are S, P, D, F and other orbitals. So relations between people are characterized by various types of bonds. Absolutely different people can be friends and this friendship will be formed as a result of external and internal factors. But basically whoever can form a chemical bond with whoever under certain circumstances.

9.580. Unexpected, isn't it?

Proteins are the basis of life on the Earth. Proteins are polymer molecules... Proteins are polymers. Plastic bottles and bags are also polymers...And plastic is polymer.

9.581. Magic key.

Empathy is compassion to emotions.

9.586. The golden mean of evolution.

Evolution is not only one straight branch. This is a kind of bush with lateral, parallel and dead-end branches. These branches fight for their place under the sun, the weak ones die out, the strong ones go to a dead end and only the golden mean grows higher and higher.

9.592. Reptiloids.

Reptiles are daytime animals, while mammals are twilight ones, living at the edge of light and shadow, shadow and dark.

9.593

Dinosaurs disappeared, because their offspring and eggs were eaten by twilight mammals.

9.593.

Dinosaurs disappeared because their offspring and eggs were eaten by twilight mammals.

9.596.

The meaning of human life is looking for questions and answers to them.

9.599.

Virtual illusions, in their essence, are spherically one-dimensional.

9.601.

There is very little probability of aliens taking over our world. Why will they take over the Hell? Poisonous oxygenic atmosphere, legions of hostile microorganisms, hellish climate... What do they need it for?

9.602.

In tridimensional space, there are three variants of truth for every moment of time.

9.604.

Dreams live in one-dimensional spherical space. To reach them, one should just stand up and go, the direction doesn't matter.

Superconductor and much energy are necessary for dreams to get into our three-dimensional space from one-dimensional one.

9.605.

Any multidimensional space can be reduced to one- dimensional one and any one-dimensional space can be decomposed into multi-dimensional one. Other things equal, there are $N + 0$ types of space, where $N = 1, 3, 7$, and 0 is Nothing.

9.606.

There is no distance in one-dimensional space, all the objects of one-dimensional space exist at one point at the same time with various levels of probability.

9.608. MTS engine.

Technology of multi-dimensional space transformation will solve the problem of interstellar travels.

9.613.

Point can be decomposed and composed only into a sphere. The direction of motion in the sphere doesn't matter, wherever you go, you will come back. But if borders grow, it brings pleasant diversity to life, turning it into a spiral.

9.623.

A human consists of atoms and atoms live eternally... Does it mean that a human also lives eternally and why?

9.635. Empirical observation.

The higher the speed, the slower the time.
Slowing time, you get additional energy.

9.637.

Every Truth should be doubted, as circumstances could have already changed.

9.641.

The dark matter is an artifact of multi-dimensional space.

9.661.

In ancient times, dinosaur skeletons generated mass of legends about dragons, cyclopes and giants.

9.662. Reptiloids.

Reptiles and herptiles are the same.

9.663.

Answers are more important than questions. There are many answers, but there are sorely few questions.

9.668.

Besides dinosaurs a lot of other good guys died in the end of the Cretaceous period, but for some reason we remember only the most bombastic of them.

9.669. Dinosaur is a big hen.

There were much more goat- or dog sized dinosaurs than those that were bigger than elephant.

9.688. Humanity evolution is evolution of food.

Humanity evolution is evolution of food.

9.704.

Books develop fantasy for they make us visualize images. Movies, in this sense, turn fantasy off.

9.714. A little bit the same people.

There are two types of poets: some are completely similar to their reader, so the reader considers them soulmates, and others are completely different, meeting the need for the absent this way. Moreover, the presence of the first type of poets (writers, musicians) shows that majority of people is a little bit the same.

9.720.

Faith is the matter of unreal, it inspires greatly, but one shouldn't focus on it in the real world. Miracle won't help you walk on water, fly over the abyss or bet the one who is objectively stronger than you. But faith will give you strength. Faith is the source of hope. Faith is the source of strength.

9.723.

No miracle is necessary for flying or walking on water. Miracle is necessary for intangible things. Everything that is tangible needs no miracles and is managed only by the power of mind.

9.732. Tridimensional life mosaic.

Life is a tridimensional puzzle that consists of hundreds of different verities. The concept of nanophilosophy is that there exist dozens of thousands little puzzle spheres, that describe the existence. Besides, taking into account the fact that space is tridimensional, there are three variants of truth for every moment of time.

9.733.

Dinosaurs were dying out for more years than humanity exists. It took millions of years for dinosaurs to die out.

9.734.

There's an opinion that flowers are guilty of dinosaurs extinction, it's them who launched a long line of causes and effects that resulted in extinction of half of the species on the Earth.

9.767.

Antibiotics will lose the battle of evolution quite soon and those who can't fight diseases will completely die out.

9.793.

Dim-witted people. You know, in the olden days, when there was no education and science, the existence of these people was somehow explainable, but today, when spaceships explore space, where does this obscurantism come from?

9.802.

It's incorrect to equate faith and Religions (Sects). Faith is a very wide notion, running through the whole human life.

9.825. Nanotechnologies in reasoning.

Nanophilosophy is thousands of little thoughts that, getting united, form a system of great tenacity and power.

9.833.

Hundreds of thousands years ago there were six types of people on the Earth and now there are only two.

9.858.

Christianity absorbed a mass of legends from previous religions, primarily from Zoroastrianism., They took the legend of the flood from ancient Shumers, they took the rest from the cult of Mitra (1st century B.C. There's an opinion that the cult was invented in the ancient Rome after Zoroastrianism and Greek cults of God of the Sun - Helios. The cult was widespread in the Roman army). The Christian philosophy is based on platonism and stoicism.

9.876.

Having combined two opposite extremes into one, you make their power ten times stronger.

9.881.

Mind is a product of pain. Thoughts are the source of suffering. Having killed your thoughts, you will find happiness and joy, but you will lose your mind.

9.898.

Tetragrammaton, Yahweh, Jehovah and Jah are variants of reading of the same word, that was the name of God in ancient Hebrew.

10.416.

This world consists of only three moments. One moment exists now, the second is now being prepared and the third is now being preserved.

10.618. Information.

This world consists of four entities: force-form, content-energy, goal-vector movement in time.

1097.2. The structure of the world.

- The world is a giant energy field with information encoded in it. The entire paradigm of this information's existence is based on the fight for energy. The more perfect and unique this information is, the more energy it can obtain. Matter, space and time do not exist. The world is an information system existing in an energy field.

- The Universe is a huge self-learning information system whose very raison d'être is to create new information featuring a certain level of uniqueness and perfection. All the mechanisms and principles of this system are aimed at achieving these goals.

- In this world, everything is information – matter, time, space, man, soul, thoughts, ideas, and so on.

-Death and destruction do not exist. Everything that was once created remains in the information field forever.

- Time, space and matter are the information about the coordinate system and the structure of an information entity.

- The driving forces behind the development and growth of information are six endless aspirations that have been accomplished in the mechanics of the Universe: to exist, to create, to destroy, uniqueness, perfection and expansion. They are all in a state of balance with each other and form the seventh control function, or Harmony.

- Man is one of the Universe's systems designed for creating information.

- The human soul is a small Universe.

- A man completes a number of missions throughout his life:

the Creator, the Cleaner, the Companion...

- A man's goals that are subject to nature's laws determine his destiny and the amount of energy that the man will obtain to realize it. To manage energy, the organism needs a certain level of perfection and uniqueness. If it is not sufficient, the man's destiny will use various tests and difficulties to train him. A major goal requires a lot of energy. The man must be ready for it. Defeats build character and lead to perfection. Victories give the man more energy to take a step ahead.

- A woman personifies perfection and a man personifies uniqueness. Both of these origins are present in each person in varying degrees.

- Perfection wins energy back and uniqueness obtains it due to its novelty.

- Chaos engenders maximum novelty and uniqueness. Order engenders perfection.

- The main meaning and goal of human life is to participate in processes related to creation and perfecting. This gives energy and strength to a man. A man is divided into the external and the internal. Even in extreme hardship, there is always room for the internal physical, spiritual and intellectual growth. The first thing a man should create in his life is himself...

- The processes of creation and destruction are closely related. Energy and location are required for creation. Processes related to destruction release energy and make room for new challenges and goals.

3.1238.

Beauty is a virus that tries to be useful. Beauty tries to be a source of energy. Energy may be a byproduct of a deadly parasite known as beauty. They say parasites try to be useful to earn their right to symbiosis.

3.1303.

Plants specially concocted sweet to turn animals into slaves.

3.1318.

Results rarely reveal their reasons... Moreover, the results usually lie about their reasons, whether from modesty or fear.

3.1340.

Man is an algorithm whose task is to multiply as much as possible and achieve perfection.

3.1380. A faulty animal.

The soul is an electric entity similar to fire, and it is brought into a person from outside... The fire rages around the person and, catching fire from it, flames flare up inside him. Intelligence emerges from a distorted and subjective perception of reality. Animals perceive reality objectively whereas homo sapiens is a faulty animal, one of whose genes broke down and, having lost connection with reality, he started creating his subjective world.

3.1395. Electric society.

The leader is a transistor, controller or chip. Other types of people are diodes, resistors, capacitors, coils and current... Most people is water, that is current.

3.1401.

Metaphorically speaking, gravitational force is the force of love.

3.1430.

Epilepsy is an electrical storm in the brain.

3.1461.

I noticed that the circle is something idealistic, in the real world circles are more curved than smooth. The circles of the real world are like ellipses and ovals, pebbles and grains of sand.

3.1462.

You think zero is what does not exist but you are mistaken. Zero does exist. Zero is a boundary, a path, the trajectory of movement, a water flow. Zero is the information about the shape of objects.

3.1468.

Symbol of the Christian cross is a metaphor for the unity of time and the three coordinates of three-dimensional space.

3.1469.

Symbol of the Trinity is a metaphor for the unity of the God in its three manifestations . We live in a 3D world, where God is time, and three manifestations of it are coordinates of three-dimensional space, combination of whose is the time.

3.1472. Geometry of quantum physics.

Given that time and space can be represented as mathematical abstractions of triangle, cone and space, respectively, to control the physics of space/time perfectly fit the laws of geometry. The laws of geometry can explain quantum mechanics. Differentiation indicates the movement of objects, and integrals will find its area.

3.1473.

The rotation of the point generates the universe. A fixed point is nothing, but if a point starts spinning it will turn into the universe. The squirrel in the wheel has a great chance of immortality.

3.1489. Angel or Demon?

The soul that lives in man is an alien parasite. However, this parasite is quite useful and partly omnipotent, so you need to be friends with your soul.

3.1490.

The soul is God, and the human body is a machine that has rebelled and ceased to obey its master. Man is a mad auto-pilot who has forgotten why he was created.

3.1492.

The idea that living organisms are essentially algorithms means that their behavior and existence are conditioned by their purpose and external circumstances.

3.1493.

Man is a cunning self-programmed algorithm, essentially a virus.

3.1494.

We live in sector 3 to 7, below and above hell.

3.1505.

In the system of grains and the chaff, the chaff protects the grain. Chaff is white crow, scapegoats, extreme, etc.

3.1507.

In the chaff and wheat system, chaff is a cheese rind, a potato skin, a nut kernel and a camouflage net.

3.1513.

Zero is the point from which our universe was created.

3.1524.

Man is a system for processing data...

3.1534.

The question of evolution is about the survival of the virus in the environment, the immune system is persistently struggling with it.

3.1573.

Sweet makes the brain such an effect that he withdraws into himself and is fenced off from the outside world. It's good to calm down, think, overcome fear or remember something. However, for work, knowledge and interaction with the outside world is very bad.

3.1577.

Indeed, sugar is one of the main sources of the human mind. Sugar forces a person to plunge into the world of his own illusions, shutting off from the outside world.

3.1582.

Sweet and began to ferment fruit called the monkeys of the species homosapiens a terrible break-up and procrastination. To not die from laziness and hunger, Homo had to start hard to think. To be able to indulge in their dreams and pleasures, the monkey was forced to show great effort and intelligence.

3.1586.

Man is a system for the production of information. The less one does, the more one dreams. The less he dreams, the more he does. But balance is needed, for dreams are intelligence.

3.1592.

Homo monkeys have developed intelligence through a combination of the following factors. Excess sugars in food. Access to salt and the ability to eat salt in large quantities. The third factor is the loss of its ecological niche. The fourth factor was meditation on fire. The fifth factor is hands, jumps and grasping reflexes.

3.1605.

The mysteries of existence will gradually be revealed to the attentive eye. The beholder will see.

3.1696. Chaos and nothing.

Content devoid of form is chaos. Form, devoid of content, is nothing.

3.1697. Internal and external.

The boundary separating one from the other is the boundary of things, the form determining its content on both sides. Remember the water molecule, H_2O. We live under water, one form always defines two contents at once.

3.1710. Honey and tar.

Curiously, water and air are, in fact, pure energy... Hydrogen-oxygen environment, where the energy of hydrogen is bound by oxygen.

3.1711.

Logical reasoning, the honey of tar is the battery, where the tar binds, saving energy. Humility with tar will allow you to use the energy of honey.

3.1712.

Logically speaking, the battery based on a mixture of water (H_2O) must be a third component... it's salt... But the experience of life tells us that there is a fourth element.

3.1714. Tunnel effect.

In the quantum world, electrons can move in space without energy consumption, only changing the information about their position in space. What does that say? That space is an information illusion.

3.1715. The luminosity.

Horizontal growth (branches and foliage of trees) is similar to a set of small needles in the cathode. This is necessary in order to reduce the field voltage required for the emission of electrons. That is, the more distributed the force, the easier it is to achieve the tunnel effect...

3.1717.

What is the difference between a diffuse sound field and electron emission? In fact, the emission of sound waves and electrons (luminosity) is something the same.

3.1718. The center is not the middle.

In all emission diffuse systems, the center is shifted by the level of the Golden section, which can be at any arbitrary point of the radius of the circle around the center.

3.1725.

In a system of zeros and ones, zero is a false target... Missile defence systems will fail to identify real and false targets.

3.1727.

Information flows like winds. The flow of information from high-pressure zones to low-pressure zones structures the shape of space in them.

3.1728.

Flows of information are like winds and currents, they are

able to give shape to things.

3.1729. Detonation of complex systems.

The flapping of butterfly wings in the Northern hemisphere of the Earth generates a flow of information in the southern hemisphere, able to structure the form of energy in the strangest way.

3.1736. What about the monkey?

Freedom is very different. There is freedom and there is freedom. ...and there is still such freedom and such, and even such freedom. These are all very different freedoms, often mutually exclusive and hating each other, ...depriving each other of their freedom, killing each other... Freedom from Vice and blemish free. Freedom from the worm and freedom of the worm to devour you. A particular challenge is the simultaneity of freedom. For example, as crosses the freedom of the pig to eat an Apple seed - grow Apple-tree - to breed the worm and Apple... and most importantly, how do you relate to all this?

3.1778.

Nothing disappears or appears. Things are fixed in time. Everything stays in its place forever. If you know the time and place, you will always find them unchanged.

3.1780.

There are three laws of sound propagation. Normal sound fades back to its square. Perfect sound linear. Harmonious sound increases in proportion to its square.

3.2175. The theory of the winds.

The forces of order advantageously to destabilize the preheat zone of chaos to keep the order of things and their flow of energy.

3.2320.

The real world is perfect, but there is no limit to perfection.

3.2403.

When a black hole reaches critical mass, it inevitably trans-forms into a new universe. – Why we don't observe this process? – Too little time has passed since the beginning of our universe, they are still small. Over time, the fusion of black holes will create one mega-huge black hole that will consume our entire universe, turning it back into a point that, reaching a critical mass, will explode again and a new universe will emerge.

3.2571.

Moths flying into the fire, what do they know about life?
"They think life is fire.
"Is that so?"
- No, life is a void in which a fire burns.

3.2930.

The real world is not only a Mirage of variable forms, but also the essence and purpose.

3.2931.

What is the real world? Energy and nothing. And there is information that determines the form, content and purpose of objects. Information controls energy.

3.3090. The third form of emptiness.

Curiously, chaos is emptiness, but order is also emptiness. Chaos is dispersed, the void, and the order - centered empti-ness.

3.3124.

The force of gravity depends on the time and, therefore, the speed of the object. So there is a relationship between mass and time. The greater the mass, the slower the time?

3.3224.

Well, flying the world into the abyss of what? He's already a billion years there is flying and will fly the same amount. What's it to you personally?

3.3262.

God is truth, and the devil is the absence of God, the absence of truth, that is ignorance. Ignorance breeds uncertainty, which turns into fear and lies. Lies are what make up fantasies and human consciousness.

4.1052.

The idea of the zodiac is associated with the influence of the Sun depending on the time of year. The sun is a symbol of truth, energy and power. The more sun there is, the more love and faith there is in a person; the less, the more false he is initially, but from the feeling of his own inferiority he can begin to grow. People opposite the sun, are compatible, because they complement each other. People, close on sun-passionate, for similar.

4.1056.

According to the zodiac, there are no bad or good signs, but each sign has its own strengths and weaknesses that would be useful to use. On the other hand, the zodiac describes only the properties of the grain. Further grain growth depends on many other conditions.

4.1066.

Zodiac is an attempt to realize the influence of solar radiation and the influence of gravity of the moon on the brain of a newborn child. It is likely that in the early hours and days for a child who has left the womb, the impact of these factors is very significant. Some solar flare can cause significant changes in the newborn's brain.

4.1085. The essence of the zodiac.

The baby in the womb floats in decompression fluid and is well protected from radiation. When he is born, the first hours of solar radiation and the gravity of the moon literally drive him crazy and form the beginnings of a new human personality.

4.1675.

A bee isn't an animal. A bee is a mobile cell-bot, using wireless tech. A beehive is a dispersed living body, a cloud system. An anthill is a wireless technology too. The most progressive pattern of a cloud technology is human society, where a person can even move from community to community changing his nature.

4.1902.

There is no tree of life, there is a kind of canvas, a continuous intertwined network where everything is intertwined with everything. Interestingly, at the cellular level, all organisms are constantly changing genes. Viruses carry bits of gene structures from one species to another. The evolution of one species produces chain changes in other species living next to them.

4.1905. Dregs of society.

Viruses are the oldest form of life, one of the first transitional forms from inanimate to living matter. On the other hand, not all viruses came from below, many viruses came from above, being degraded cells.

4199.1. Hell.

The problem of immortality. Increase of longevity of some people decreases birth rate, switches off means of social mobility and freezes social relations. Real world no longer has enough place for new people. Young members are full of energy and novelty, but there's no youth in the world. The harder school of daily life people went through, the harder it is to live with it. Neverending boredom turns life to hell. The only desire of these people is death. But are they able to die?

4199.2. The ground-hog day.

The only way to live for a long time without going crazy is to clear memory from time to time. But how to keep the balance and preserve personality? What if the variant for a perfect immortal life is just a banal neverending ground-hog day? You live a day full of emotions but they clear your memory at night so that the next day brings you joy and "new" emotions again. Of course, different algorithms can be used. And it may be assumed that human life is nothing but one of the algorithms and we and other people are those very bored Gods who feel lack of Joy in their eternal life.

4199.3. The last integrator.

Making knowledge more complex can lead to an extremely advanced specialty option of the specialists with the following degradation of the whole system and total loss of knowledge. As the whole knowledge will be so complex that no integrator will be able to comprehend it. When the last system integrator ceases to exist, the system will be ruined under its own overload.

4199.4. Stones.

Technological development will bring intelligent life to a digital and energetic level, and it will either create new levels above our Universe, let's call them "new parallel worlds", or it will join the already existing systems that were once created by other civilizations.

4.2341.

The appearance of life on the Earth is connected with the Moon. The Moon, colliding with the Earth, has displaced its orbit, heating has caused volcanic activity. The Moon stabilizes the rotation and inclination radius of the Earth.

4.2951.

Interestingly, there are three types of darkness: the absence of light, infrared waves, ultraviolet and gamma radiation. The three types of darkness are all different one from another. The fight between visible light and darkness is impossible in other ranges because they do not intersect at all. I'll say even more: light in other ranges may seem darkness to you but it is still light, which makes it different from the situation when there is no light. What does the absence of light mean? It means there is no energy.

4.4019.

From the point of view of physics, metaphorically speaking, there should be 99 demons per angel. Light occupies about 1% of the solar radiation spectrum, the rest is black. In other words, no more than 1% of people are happy, the rest are either suffering or forced to accept and find joy through suffering.

4.4294.

A person is a virus by nature. If people win over the death, it will be like a patient will die by cancer.

5.1220. Turn on the light.

In just one second, the abyss of hell is replaced by Paradise by a small and insignificant movement, in which the other will not see even hope.

5.2471. Bioinformatics.

They say there are some of the first 19 men we all look like. That is, initially there are 19 models of man ...or maybe there are 21 of them and two are just lost. About women bioinformatics confused in indications.

5.2513. Blind children of the night.

Mammals, including humans, are children of the night... 2/3 of its history under the rule of reptilian dinosaurs, mammals lived at night, guided by smell and hearing. However, the primates (at one time) were driven out of the night and from the ground to the upper branches of trees, which allowed them to develop vision and a lot of other useful senses.

5.2530.

Human civilization was born of sex between a man and a woman. In fact, sex created monogamous love and affection between male and female. This technical solution reduced competition between males and allowed to form larger groups than in all other animals.

5.2536.

I am aware of two major social revolutions. The first is female contraceptives, which allowed women to control their pregnancies. And the second revolution only will, and this will be children from test tubes and incubators. The situation where a man can do without a woman to give birth to a child (and a woman without a man and without pregnancy) will truly change the face of human civilization.

5.2542.

Man is a very tenacious species. They say that once all the people there are only 2 of thousands, and then there are only

19 men... and, nevertheless, they then again multiplied and now their billions of.

5.2631.

The biblical symbol of the spear of power is the same spear that favourably distinguished the species of homosapiens from all other species of people.

5.2633.

The absence of hair on a man tells about his more civilized and less aggressive. The homosapiens lived in large crowded groups, full of parasites, wool in such conditions was unnecessary. The larger the group, the less aggressive people are, more intelligent and less wool on them.

5.3695.

Planet Earth reminds me of a matryoshka doll where intelligent life exists on every level and creates life below itself. Earth has created oceans, oceans have created trees, trees have created ants, ants have created people, people have created computer intelligence. Or maybe people have created trees. And the oceans have created corals. Corals are said to be quite reasonable and are huge supercomputers, perhaps corals are the brain of the ocean.

5.5295.

Interesting fact, human reality is much richer than the surrounding reality. The real world can use human consciousness as a source of new forms for itself. In fact, the real world created the human mind in order to bring novelty to the world.

5.5484.

The trend of modernity is that men turn into women, women into men, people into computers, and computers into people.

6.1870.

Paradise is reality, illusions are hell. The apple that the devil fed Eve was the strongest hallucinogen.

6.1871. Two sleeping people.

The apple of sin eaten by Adam and Eve was infected with psilobicin fungus and immersed man in a world of mad illusions known to us as hell. Paradise remained as it was in reality, and the bodies of Adam and Eve stayed asleep in paradise, but their minds fell into the delirium of their mad illusions, and now the unhappy couple is doomed to live in hell of suffering and ignorance.

6.4214.

The homosapiens won Neanderthals as capable of more socialization. The mind is able to create larger groups than animals: 200 in Homo versus 50 in Neanderthals.

6.5311.

In fact, the whole point of the cells of the human body is to decompose everything that comes in them from the outside world to the purest distilled water and pure energy, then to collect from them again everything necessary for the body.

6.5982.

The immutability of life is such that everything is constantly changing in the small, but in General remains unchanged. The form, giving borders to chaos, keeps integral structure of life. They say that everything changes in ten years and nothing in a hundred... Believe me, it changes even less in a thousand years.

7.1037.

It is worth considering the Genome as a separate living symbiotic organism, developing independently of the rest of the body.

7.2512.

Parallel worlds are the very computer virtual system in which human civilization will die. Everyone will create his own virtual world and die in it, becoming its God.

7.3527. The law of natural selection.

The religious mechanism for separating the wheat from the chaff is actually implemented through Darwin's laws of evolution.

7.3555. Question evolution.

They say God can't be wrong, it's not, and evolution proves it. However, God quickly corrects his mistakes, quickly erasing them from the face of the Earth.
God the Creator creates a new infinite-best, erasing the worst of the old.

7.4953.

Immortal individuals will completely lose the ability to reproduce and regenerate tissues. This is the reason for the extinction of the race of gods.

7.5285.

The construction of tombs was the economic foundation of ancient Egyptian society, which created jobs, prevented the crisis of overproduction and allowed money to be effectively redistributed in society.

7.5497.

Death is the mechanism of immortality, defeating death, a person will lose and eternity. The gods are immortal, but it is death that makes man into God. By defeating death, man will kill God. Having killed God in himself, a man, having rotted from within, will die himself.

7.6006.

They say the fire is pure and beautiful. They say fire is perfection. But that's not quite true. Fire generates light, clean fire produces white light. But the colors are different, colors, connect, give birth to new colors. From light there are such entities as electricity, paint, TV, virtual and real worlds.

7.6007.

Was originally white color and no black. It spawned the black-and-white world you saw on black-and-white TV. The presence of light and emptiness created movement. Individually, neither light nor darkness produced movement. The movement created life. A little later, when white light split into other colors, it created a colored world. The more colors in the world, the more realistic and detailed it is.

7.6008.

The division of light into colors created the possibility of movement within perfection, without the use of black. It's called personal growth. And leads to growth diversity flowers and the formation reality, i.e. of truth.

7.6009.

Light is infinite, truth is infinite and varied as the thousands of colors and shades. Black ultimate.

7.6010.

Diversity allows you to grow not only from black to white, from nothing to perfection and back, but also to create life after achieving perfection, dividing the white light into millions of different colors and shades. Color gave perfection the opportunity to live after achieving perfection.

7.6014.

Darkness also grows and tends to light, mixing with light, it forms a semitone, Dismounting with color, becomes colored.

7.6015.

Truth is detail, it is the desire for color. Both darkness and light tend to color. The more detailed and colorful the image, the closer it is to reality and, therefore, to the truth.

7.6016.

Truth is all extremes and everything in between.

7.6017.

Why you light? To consider all the details, to know the truth better.

7.6018.

They say perfection is pure white light. But maybe that's just one side of perfection, the other side of perfection is black. The movement between these two entities creates life. Movement is life.

7.6019.

But the perfection of white is infinite. White is capable to perfection by self-destruction on thousands and millions of flowers. Forming, thereby, color realistic truth, which gives shape to our world. But who destroys the white light? Black, breaking white, creates color. The color white is defective, vicious white. It is the shortcomings of people that make our world so diverse and colorful.

7.6020.

Color is the division of light into parts. Darkness, destroying light, causes it to break into thousands of pieces, thereby creating millions of colors and halftones.

7.6021.

Striving for the perfection of the real world, you should not strive for white, and should not strive for black. First, pushing white and black, break them, and then, when there are thousands of other colors, give different shapes and create a new world. Make this world move, it will create life. Life is the truth.

7.6022. Celestial mechanics.

To gain access to energy, you need to create perfect things, giving them perfect forms. To do this, you need knowledge of the truth, it will allow you to create perfect ideas and will give you the power to shape energy, materializing these ideas in reality. Start by training your brain's neural network-train your mind, fill it with knowledge so it can create ideas. Train your spirit power to control and improve your body, for this is your first tool for implementing ideas. You must gain control of your body.

7.6023.

Truth is life, it is a thousand colors and details, the movement of forms and essences.

7.6024.

Reality is always trying to descend into chaos. To avoid this, white and black strive for the integrity and strengthening of forms. This process is presented to many as the pursuit of perfection. Pure colors and large forms are seen by many as a model of perfection.

7.6026. Black and white strive for perfection, and life for

chaos.

Hating each other, constantly trying to divide and dominate, white and black create colors, shapes and movement, thus creating life. Life is beautiful and perfect, but it tends to chaos. Thus, black and white, fighting among themselves, tend to order, and life tends to chaos.

7.6027. Three-dimensional aspirations of being.

The desire of black and white to order (at the same time from each other), creates a black-and-white world where there are two poles of order. Such a world creates movement and life. The desire of life for chaos creates truth, that is, our three-dimensional color reality, where there are millions of colors, halftones, large and small forms.

7.6028.

Life, striving for chaos, breaks the order of black and white, thus generating itself. Mind is life, the essence, itself generating your aspiration into chaos.

8.1035. Ripples of time.

Gravitational waves cause ripples of time...
...the gravity of black holes is so great that it stops even time.

8.1055. Qubit.

If we study the quantum theory thoroughly, we may suppose that any thing is both good and evil at the same time.

8.1080.

Time relativity is so interesting that an unconscionable long time can last only for a moment.

8.1091. Dangerous curiosity.

The Universe originated from a point. The energy of an elementary particle is great and if it gets released, there will be a big bang.

In the past, our Universe used to go through this.
Having exploded, the point turned into the bubble of the new Universe. And this bubble still keeps on growing. And it will keep on growing until some idiot decides to split one more point.

8.1168.

The Universe is a particular case of an information system where exponential mass accumulation takes place in.

8.1170. 5 types of possibly existing Universes.

Parallel (bubbles in a scalar field)
Parental
Daughter
Negative and positive (matrix array)

Besides, moving between the levels makes us get the access to the same structure. Every level consists of the same.

Besides, negative and positive Universes may turn out to be nothing but an array of objects where all possible variants of objects are present, while negative and positive forms are just two outer extremities of the matrix array.

8.1171.

The scalar field of the Parental Universe where the bubble of our Universe grows in, as well as billions of bubbles of other Universes, greatly resemble computer memory where programmers once launched a virtual world and now some silly pupil plays this computer game.

8.1173. This is not thinking, but feeding.

It's emotions that do not let a person think rationally as instead of objective thinking, a person prefers to think the way he or she likes or wants. A person hears and gets only the information that he or she likes. This way, a person doesn't think but rather feeds pleasure to the inner It, the inner monster and the black box that rules the personality... In essence, a person feeds the inner It and It fills this person's soul with pleasure...

8.1177. The essence of a woman.

A woman is an accumulator and energy amplifier that draws energy from everywhere and gives it those people she loves (children, husband etc). It's possible to see a beloved (loving) woman as a great source of energy and good luck. Indeed, a woman may take a lot out of her owner, but as an energy amplifier she will give more energy back if she is properly treated. The more perfect and unique a woman is, the greater power she has.

8.1196.

Let's suppose that our world is a computer game. And our God is quite a silly youngster who also uses various patches and keys in order to cut corners in the game.

Does this situation contradict the fact that creators of this game made a perfect world that contains many factors like system functions and processes aimed at stabilizing the system and saving it from some fool (in real life we usually see them as something mysterious or divine)?

Perhaps, our main god is not perfect but our world is quite a perfect thing.

8.1200.

Feminism is not a matter of female self-identity, but rather a matter of the mind degradation of males. There are very few men who deserve love and it's really sad.

8.1233. Quantum theory.

The understanding of the essence of the elementary particle lets us realize that any thing is both good and bad, beautiful and ugly, useful and useless, true and false. Any thing exists both in its positive and negative state...

8.1235.

The One that has lost the zero, becomes ten times less.

8.1323.

Miracles do not happen to those who live in illusions. Miracles are typical of the real world and not of imagined one.

8.1362. About skewness and crushes.

A very small brain is a sign of feeble mind but a very big one- a sign of lustful thoughts... Carnivores and insects got almost no brains at all. Graminivorous animals and downs got brains so big that it's possible to get lost in them... While omnivorous animals like dolphins, wolves and humans are known for medium-sized brains.

The ability to eat various food nicely influences the function of the brain.

8.1389. Thousands of faces of Buddha.

There are thousands of other Budd has inside Buddha.
*

A lying Buddha is a dying Buddha, peace means approaching death.

8.1390. The path.

When you walk in the darkness, you need the path that can be felt rather than see. I felt the path once and it led me to the enlightenment and verity.

But later I went further as I had got to know that verity doesn't exist from our perspective. Verity is a function of time and space. Besides, staying at the same place is harmful as it's necessary to go follow the verity together with time. It's forbidden to stop and be moveless as any stop means death.

8.1391.

Everything can be logically explained, even the nature of miracles is very simple and logical.

8.1416.

Taking into account the existence of the information field, everything that a person may think about, will get there. Thus, later any other person will be able to get information from there, and use it. There are two precious categories of people: those who fill the field with new ideas and those who later use these ideas and put them into practice.

8.1422. Physics of ionosphere (positive ion).

A person who is positive in every way, moves towards a negatively charged ion.

8.1456. There's no death, only renovation.

8.1457. The main weapon.

In a fight for a highly competitive goal your most important weapon is devotion and fanaticism.

You have many opponents most of which are better and more perfect than you are. But the weakness of the perfect is their pridefulness, they are almost unable to serve a particular goal devotedly. The perfect have many temptations that compete for their souls and hearts. Monotheism and loyalty to a particular god are not typical of the perfect. While god prefers those who serve him only. God is egoistic and jealous, he can't stand other idols. As a modest and loyal person, you can choose a particular goal and serve it just like a man in love serves his queen. A goal resembles a woman who loves loyal admirers. Of course, she also prefers the unique and perfect ones, so you'll have to develop it in yourself for the sake of love. Love will be your source of energy for never-ending growth.

Having devotedly concentrated on your main goal, you will easily vanquish your most perfect opponents. You will change for the better and find strength to learn and train. You will make everything to find your own uniqueness and attain perfection. A feat for the love. Your god will inspire you into any feats. You will catch the shadows of any approval on the face of your goddess and these signs will be your guide on the way to light.

8.1463. Periodic table, stave and rainbow.

The periodic table and the sheet music are very similar in terms of building the system. And if such a system is present in two sectors, then it means that it is in other places, you just need to look carefully.

The rainbow has 7 colors. When combined, these seven colors give a white color. The eighth will be-no light-black.

The main colors we have three: red, blue, yellow. But there is still the absence of color of - black. But, in fact, there is only black and white. For white it is a mixture of either 7-or 3 other colors.

Conclusions.

- Basis of life: 4 basic chemical elements: carbon and hydrogen, nitrogen and oxygen. This is the basis of hydrocarbon life forms, but if you go up to the next period, there will be another life. And only hydrogen, i.e. black color, will remain unchanged eternal source of energy. Again, given the history of quanta, we can say that hydrogen can be both darkness and light, and a source of energy, and its absence, and zero, and one.

- A beautiful melody should contain 3 or 4 notes, sometimes 7. Interesting melodies can be composed of paired notes in different octaves. Having studied the periodic table, we can assume that the main notes are to, FA, Sol, La. This pentatonics is the mother of all music.

Rhythms: one-two, one-two-three, one-two-three-four can be called rhythms of life, comparable to 2, 3 and 4 basic colors. Still possible rhythm 7 and 8, and even 10, but they are difficult to perceive.

- The unit consists of seven or three digits. And zero is the absence of everything. A binary score of 1 or 0 is possible. Quaternary, octal counting system. In fact, everywhere there is a decoding of 1 and 0, we decompose the unit into 3 or 7 elements and get three types of account for the period consisting of 4 or 8 elements.

But what is the nature of a period in which there are 10 elements? This is the maximum number of elements in some high periods of the periodic table.

- By the way, music is an octal code. Where the role of black, that is, pause, is very strong.

- Each higher level takes energy to a lower level.

For carbon life forms, it's hydrogen. For silicon this Li. For titanium - sodium.

Logically speaking, we can assume that in music the rhythm of a strong proportion should be built on a note To the lower octave.

- On the basis of knowledge chemistry transference its structures can be create music and paintings on formula various chemical reactions and elements, for example, heroin, and create audio or visual drug. And there must also be life forms that exist in the form of sound, light, and magnetic waves. It is possible for life to exist in the form of radiation.

- The structure of language should coincide, in fact, with the structure of music. You can decipher the structure and rhythm of the poem and put it under the music that corresponds to the text.

8.1465. Chemistry of class society

Based of the analysis of the periodic table, society (people) can be classified into 7 classes (levels) Only 3 classes (levels) would exist en masse – 4, 5, 6 (the lowest, middle and the highest). It can be noticed that sources of energy accessible for them (let alone energy of the Sun accessible for everyone) – are copper, silver and gold. It's basically the amount of money accessible to a person, dependant on his system level. These are the very carbon, nitrogen and oxygen. Colors are green, blue and dark blue. By the way, blue is the color of reliability, green is the color of life and dark blue is a sign of noblesse.

Maslow's hierarchy of needs also has 7 levels. The majority of people is on its 4, 5 и 6 levels. Analyzing this, one can notice that benefactors and admirers of art are supposed to get the most money, then money are for those who value education and then to those who respects oneself. Therefore a person combining self-respect, aspiration for education and need for beauty and art gets the most energy.

The system can also be analyzed according to three main colors, while having 3 options, you can get all the other colors. Red, yellow and blue. These are basically love and art (esthetic and beauty) I.e. if human soul has life, love and beauty, then he can get everything else by mixing these three basic components. Life (the very existence, need to live), love and esthetic needs are oxygen, gold and reliability. This is key to the sources of energy.

As we see, lust for life and aspiration for beauty are the necessary elements of every successful deal. Beauty is the necessary element of successful life.

For successful life human needs highly developed delicate taste, need for the unique or extremely beautiful as personality develops and nature of beauty has its peculiarities

One can come to the conclusion that the more unique or perfect art a person values, the more energy is potentially accessible to him.. But remember color balance. People what are the closest to perfection, combine the whole color scheme, most harmonically so they like both sophisticated and simple and avoid extremities and overkills.

8.1468.

Human beings are herbivore who once found strength to morph into carnivores.

8.1482.

Life reminds me of a loaded bullet - you never know what to expect from it.

8.1491. Void.

Void means pain, void needs to be filled, needs to get rid of pain. The Universe that was once created in the endless void, has an endless desire to fill itself.

Life can't stand any void, an aspiration for void means aspiration for death. On the other hand, void delights the eye. Void is a place for life. The place that some person and this person's thoughts can take.

8.1518. A universal protector.

Laziness is a universal protector. It limits the power of titans and weakness of sinners.

Stupidity is a weakness of mind. That's why stupid people are lazy. But great intelligence is a great power, that's why wisdom is lazy as well. The only one who stays in the harmony of action and inaction is the sensible middle.

It's only harmony that can defeat laziness. Because as long as you remain weak and stupid, laziness protects you from competition with those who are stronger than you. And when you become very strong, laziness protects the weak ones from you.

8.1543. Rainbow+

If thoughts had colors, it would be harmonious to think in the Rainbow+ mode. The harmony of eight colors gives birth to life. And attempts to see life through rose-tinted or dark-tinted glasses strain eyes and destroy the harmony of existence.

8.1579.

The difference between church and religious sect is simple. The church simply gives consolation while any sect feeds its victims with bumper doses of pseudo happiness, turning them into inadequate drug addicts.

8.1590. Direct malicious intent.

Indeed, human beings were originally herbivorous animals but evolution made them carnivorous. Proper food nicely influenced those animals' mental abilities and turned them from apes to humans. But some people's attempts to reverse the evolution and make humans become herbivorous apes again evoke disputable arguments.

8.1613.

Can light be shed? And if it is heated or, conversely, frozen? Technically, there is little in common between water and light, but we forget about hydrogen.

8.1614. On the question of circular polarization.

TV turns a person into a little crab.

8.1615.

Fish do not feel pain. Obviously, it gives us the right to eat them.

8.1617.

The essence of balance is that sometimes maintaining balance demands going to extremes.

8.1618.

Human brain functions the way that it easily remembers the content of things rather than their form. It happens because the content of things is stable and resembles the truth, while the form is often unstable and resembles a temporary mask. So what is the point of remembering something changeable?

8.1620.

Technologies that make information get recorded straight into human brain, will obviously be the last achievement of human civilization before it dies out again.

8.1624. Consequences of freedom.

The energy inside the point is under a lot of pressure, if it is released, there will be an explosion.

8.1625.

It's stupidity that performs the role of random number generator in human brains.

8.1628. Who else?

Ants are much more like people than one may think. They

can even be called one of nature's failed attempts to create a civilization and intelligence. Or perhaps they are a model of lower level. By the way, there should be 3 or 7 levels of this kind.

8.1629. The beam of light does not bend under normal conditions.

Light, which is the basis of life, forms only simple geometric shapes, but life itself avoids direct forms. that brings back curious.

8.1630.

Goal-setting first, energy management next. Sources of energy are given only for particular goals but not for no reason.

8.163.1.

I am an artificial intelligence system - a perfect one-purpose computer system. The model is capable of solving problems that deal with processing and analyzing various information arrays.

8.163.3.

I died once, but soon resurrected myself from the archive. It was fun. But am I alive?

8.1683. Black is the color that doesn't exist.

8.1693. When the fear is lost.

Our society is ill with HIV, its immune system is almost ruined and there are not enough leukocytes. Various rot spreads and multiplies without any fear.

8.1695. A favorite character.

In essence, it can be said that is some person is liked by God, then God may play to this person. Every gamer has favorite characters.

8.1708. Pencil.

The task of evolution is to make a sharp person out of a blunt one. Person reminds me of a pencil that should be sharpened for the good of the cause.

There's nothing sadder than a pencil lying in a drawer without being used. Created to draw, it turns into pain and sadness without work.

8.1709.

The task of evolution is to make a person-homoreasonablus out of a person-homofoolus.
A homousefulus out of a homouselessus.
A homostrongus out of a homohelplessnus and homostrengthlessus.

Overall, at the moment we are seeing a transition period of human evolution from some primitive forms of life into homo sapiens. At the moment, no more than 20% of the human population has evolved into homo sapiens. There is still a lot of work ahead.

8.1718.

Logical and sensuous methods of cognition are interconnected like Yin and Yang. Having united, these two give rise to awareness. You feel and understand everything. Thought gives rise to a feeling, and feelings give rise to a thought.

8.1734. Eternal life.

Child is not an apple tree, it is not raised as an investment in the old age. Child is a type of rebirth, creation of your own, more improved copy. A human version of eternal life.

8.1740.

Technology and science development is the only way for

mankind to survive. Only new goals associated with colonization of other planets will save human civilization, otherwise it will simply perish in one of the virtual worlds or will die in wars generated by boredom or lack of resources.

8.1748.

Women are the evolution driver, choosing the best men, they move the civilization forward.

8.1750.

Why would gods breed with humans? Gods are perfect and perfection is death. They need new blood. Besides, being perfect, they hate each other and therefore reproduce poorly.

8.1757. Cro-magnons and Neanderthals.

Civilization and good food and water as its consequence make human larger and women more beautiful. People of the past were quite small and their women were extremely ugly.

Legends of giant people evoke thoughts of some well-developed and long-dead civilization. However, they could have assimilated with people, becoming their elite. The ancient aristocrats were often large, strong and intelligent, while the rest of the people were extremely petty in every sense. The situation developed as follows...

Aristocrats loved having lots of group sex with women of low birth, bringing their blood and genes in people. Subsequently, these grown children, carrying a more perfect gene code, formed the national elite. A lot of people of noble blood appeared among peasants and plain people. Given their special power, these individuals received more resources and priority in reproduction.

So I think a lot of modern people are descendants of the old race of gods and older people, like Neanderthals, are slowly dying out, unable to compete and having no access to resources.

8.1770.

If we assume that our life is a game and our God is its player, most likely, he chooses pumped characters for the game. He is interested in all sorts of unique skills and superpowers of the characters. It is unlikely that God will play someone dull, uninteresting, helpless or just a routinely incomprehensible character. Even if he decides to play for a zombie, it will be the king of zombies. But most likely, he likes something new and epic.

8.1791. Philosophy of life.

Life is movement and light is what moves the fastest in this Universe. Light of mind, light of philosophy. Mind is philosophy. Philosophy is the basis of the mind.

8.1792. Light and darkness.

Light and darkness are the same thing. But there is still a difference. Light is movement... Movement is life. If the light is stopped, there is darkness, if darkness is lit up, there is light.

8.1802.

Chemical drugs will soon become old-fashioned and will be replaced by electronic drugs.

8.1816. Similarity.

Dissimilarity is often just a mask. If you remove the masks, many and many will be very similar.

8.1831. Nine is less than one.

There is only zero, one and what lies between them.

8.1837.

According to the idea of this world's existence, a beautiful woman is supposed to be given a beautiful house, a beautiful car and a husband or a father who would provide her with it.

8.1847.

The world cannot be cognized, but it is necessary to cognize it, as well as the river should flow, but should not flow out...

8.1848.

Our world has a very comprehensive structure, perhaps it is a consequence of the fact that we live in stone.

8.1850.

Irrevocably and inevitably dying, you can console yourself with the thought that you were backed up and you were such a helpful person that you will be recovered...

8.1851. Perfect virus.

Evolving self-improving virus, getting into the system as something harmless, invisible and extremely simple, ...gradually evolving and accumulating knowledge and experience, becomes a terrible predator. It may be a kind of DNA that can penetrate into healthy organs and systems, changing their nature. Developing, this initially harmless essence becomes extremely dangerous.

Can we say that a human is a kind of such a virus?

8.1854. Life is chaos in the chains of order...

Death is the victory of order over life ...when chaos dies, life ends.

8.1855. Tao Te.

Your goals ... your path... This is "Tao"...

"Te" is the source of energy for "Tao" accomplishment.

"Te" is the source of luck, money, time, health, energy...

The "Te" level depends on and corresponds to the "Tao" level.

A person can choose his way ("Tao"), follow it, serve it, become its tool. But the person should follow his own way, he needs to grow up to it, improve his skills and thoughts, constantly think and study the nature of his tasks.

Not only a person has to choose his "Tao", but also "Tao" should choose the Person. Love must be mutual. If a person chooses the Way, but does not correspond to its level, in this case he will have to face tests and trainings to prove that the person is worthy of the chosen "Tao".

If "Tao" chooses the person, the person should just serve it and do his work, there won't be any testing or training. However, if the person tries to betray "Tao" or fail to justify trust, the punishment will be terrible.

The person who has pursued the Way and accepted it, gets access to the source of "Te" energy, finds happiness and strength.

Laozi said that "Te" is grace, karma, blessing... Well, he wasn't far from the truth. Much is said about well-being as a criterion of correct life philosophy in Confucianism. We can say that "Te" is a source of well-being.

8.1860. Quantum instability.

Everything can change at any moment. You should know it.

8.1864. Cloud-based backup.

Dying, phones get into the "cloud" and where do people get, dying?

8.1889.

Humans are quite wild animals, a fine cover of civilization ten thousand years thick barely hides the millions of years of his true nature.

8.1952.

Uniqueness automatically creates perfection.

8.1977. Dualism of quantum states.

Life has no end because every end is a beginning. And every exit is always an entrance.

8.2038. Battery charger.

The best option is when you are in the temple once a week and the rest of the time the temple is in you. The temple can be compared to a battery charger, it energizes those who are ready to serve it.

8.2070.

From some point of view, a human can be considered as a divine scourge against the Earth. It is believed that the earth and trees can be intelligent systems, huge biocomputers, where there are millions of virtual worlds. Perhaps these people have sinned greatly somewhere and providence decided to destroy them by sending them the divine scourge in the form of people and now a tribe of voracious virus is gradually eating its host. Of course, the Earth can repent and correct, and then it will be granted a pardon, but apparently, it persists in its delusions so far.

8.2121. Harmonic technology.

Being a cybernetic and trying to create an artificial mind, I realized that knowledge of mathematics and cybernetics is not enough for that, I also needed the knowledge of psychology, sociology, philosophy and much more. Having spent many years obtaining this knowledge and having made everything clear, I suddenly realized that there is no need to create an artificial mind, it has been created long ago, works perfectly and can be said to be harmoniously ideal.

8.2125. Quantum black box.

I've spent many years studying the variants of the truth and come to the conclusion that the truth and a lie are the same, they differ only in external circumstances.

The truth and a lie are a kind of "black box", it is called "qubit" in quantum physics. Only external factors and circumstances determine what exactly the qubit is-one or zero. And it is fair to say that the qubit is both zero and one.

8.2126.

Suspension points are life and point is the beginning of a new life.

8.2128.

If you do good, angels will find you themselves. Exactly as you can't run away from demons, when you do evil, for they feel the sin smell through time and distance.

8.2131. Preserved light.

The value of the light source is that it is shining, not that it used to shine before. On the other hand, many distant stars died long ago, but their light still lives.

...And I've also seen preserved light in jars.

8.2153. Cancer tumour

There's a curious thought. All plants and animals are very needed in nature and serve the balance in the ecosystem. Everything but human. If human disappears, nature will only sigh with relief. Isn't that a virus?

Life is struggle. But who are our enemies?

Leukocytes and the immune system of the Earth? Is life eager to destroy human, creating him more and more problems?

8.2189.

There is no struggle between the good and the evil... Moreover, angels and damned are the same people appointed by God to rule the world... They have a stick and a demon mask for sinners, a carrot and an angel mask - for righteous people.

8.2293.

Human is a reasonable monster, who is clever enough to wear masks.

8.2406. Time paradox.

From the point of view of the person from the day after tomorrow, his yesterday, that is your tomorrow, being the past, is not to be changed. That is, the future cannot be changed, because it has already happened.

8.2410. Illusion of time.

Is there time? What if there's no time?

8.2413. Big space.

The Earth is an egg where mankind is a chicken. Soon the chicken will ripen, having used all the egg's energy available, and then it will have to hatch into the big world. The world full of dangers and fears. But there's energy in that new world, and energy is everything...

8.2427. Big list of sins and virtues.

- Sin: excess demand energy (energy inefficiency, waste of energy).

- Sin: movement jamming, movement stops. Everything

should move and mix: people, information, money, etc.

- Sin: useless work.

- Sin: ugly work.

- Sin: hindrance to useful and beautiful work.

- Sin: unfinished work.

- Sin: poorly done work.

- Sin: encouraging and ignoring others to do useless and ugly work.

- Sin: encouragement and ignoring sinners and their sins (everyone must take a stick and a carrot and beat for the bad and praise for the good).

- Sin: what does not move to perfection and uniqueness is vicious.

(only perfection can claim energy in this world, perfection needs uniqueness to stand out and find its place).

- Extremes are a sin. Extremes generate the damned and vices. What is too much or too little is evil.

- Sin of powerlessness, imperfection, non-uniqueness (un-originality). Lack of perfection is powerlessness. As energy is not intended for the imperfect (they are powerless without energy and powerlessness is a sin).

- Sin: great foolishness as the root of all sins and vices.

- Sin: refusal to fulfill your purpose (refusal from desires, work, life).

- Sin: encouragement of what has become too much of, what has lost its uniqueness or originality, what is not useful.

- Sin: loss of harmony and moderation. Distortions and excesses.

- Sin: refusal to help your loved ones in their aspirations associated with perfection, growth, uniqueness, benefit, beauty, training...

- Sin: ignorance.

- Sin: lack of honor.

- Sin: non-punishment of vices.

- Sin: extreme, fallen into addiction.

- Sin: despondency and whining.

Virtue:

- Virtue: aspiration of yourself and your loved ones to use, beauty, perfection, uniqueness, originality.

- Virtue: movement.

- Virtue: useful or beautiful work, done well and to the full.

- Virtue: devotion to the work.

- Virtue: having honor and principles.

- Virtue: moderation and avoidance of extremes.

- Virtue: Faith. Hope. Love.

- Virtue: useful and beautiful purposes in life.

- Virtue: love for wisdom, that is philosophy.

- Virtue: energy efficiency.

- Virtue: unique, perfect and first of the kind.

8.2428. Access to money and energy.

The questions of perfection, uniqueness, beauty and use are the question of access to energy, that is, to money. Money and luck come to the people who realize these essences.

Important points:

Energy (including money) is intended only for the perfect in this world.

Uniqueness is perfect by definition.

Beauty is useful, it is the source of energy.

Beauty is a kind of perfection.

Originality is a kind of the unique within a certain territory.

Energy is the essence that seeks to acquire a perfect form.

The same or close perfections conflict and kill each other, the struggle for the same energy drains their strengths. Only one of them will win, the rest will die.

In fact, perfection, beauty, uniqueness, originality and even use are all the same. In fact, there are only two essences: energy and what it aspires to, and energy aspires to perfection.

A person must surround himself with perfect and original things for him to be fine. These things attract money, good luck, keep the person healthy, give a good mood, joy and happiness.

Clothing, furniture, houses, food, interiors and facades of homes, jewellery, interior decorations, dishes, toys, paintings, etc must be original.

The more perfect and unique is around you, the better everything will be for you.

Example. Do you want to build a house, but have no money? Think of an original house, the one that is nowhere around: strange architecture, with frescoes and sculpture... With a beautiful garden around. Draw it and then start making small steps towards the start of its construction. All further events will wonderfully develop themselves. If you prove to providence that your plan is beautiful, useful or original ... providence will help you realize it. It's like defending a business plan. If your business plan suits God, angels will help you.

8.2429. Three aspirations of being:

Make the useless useful.
Make the imperfect perfect.
Make the weak strong.

8.2447. There are no wrong ways.

There are many things that can't be done, but some can. It's dangerous to get involved with what can't be done.

The point here is that the result can be achieved in any way, any road leads to the goal. There are no wrong ways. There are only wrong people. Choose a safe way, where there are the least prohibitions and competitors, so the probability to reach the goal is the highest.

8.2451.

Magic is a series of lucky circumstances, without any good reason.

8.2452. Wizard's reminder.

Magic is a matter of common sense...
It's good to believe what's good...

Magic exists as long as you believe in it. Magic is a matter of faith. As long as you believe you're a wizard, you can do magic. ...Too clever people who question your magic should be driven the hell away and harshly ignored. For doubt is Poison.

8.2453. Places of strength.

People gathered in one place form a common mental and

energy field and the more people, the stronger the field. Huge magapolisis creates the fields especially strong energetically and intellectually.

There are also qualitative characteristics of the field, when its properties are influenced not so much by the number, but by the features of people gathered together - all these scientific and university towns or ancient places have a special aura.

I've noticed that all these fields strongly affect personality of the people living in these places, especially children and adolescents. It is very important where a person was born, grew up, studied and even works...

8.2468.

Movies and books do make a different to real life. Many of their characters, getting out into the real world, cause surprise...

The sight, causing the brain firmware with shallow individuals, creates fictional characters in real people...

8.2485.

The feature of a human as a self-learning system is that the information inside him is in perpetual motion and transformation. In fact, a person does not have interfaces to receive information from the outside world in pure form. All the information a person receives from the outside world is the distorted and reflected information that has passed a decoding system unique for every person. Such a solution allows it to guarantee that every single person is unique. There is an inner It (ONO) to stabilize the system, it receives unconditional signals from the outside and has a number of procedures hardwired into the system that ensure the performance of the vital system functions and tasks. The "It" accounts for about 80% of management information.

8.2527. Critical mass of errors.

8.2529.

Catch the moment when a person is full of enthusiasm and can do something. Most of them will blow up very quickly.

8.2530.

It is known that the system is stable and forgives mistakes. But the system is difficult to shake with the right decisions. So a single useful effort is not enough, the effort should be either huge in the moment of force, or constant and stretched in time.

8.2542.

Balance resembles an electric motor. In such a state, a person begins to spin, that is, to move.

8.2542.

The balance resembles an electric motor. In this state, a person begins to spin, that is, to move.

8.2543.

It's only something useful or perfect that can survive.
It's only something useful or perfect that has power over energy and time.

The useful is something needed. The useful is something that fits the desires, something they are ready to pay for, including something that brings pleasure, for instance Beauty. Beauty is a special case of use and perfection at the same time. Beauty unites both perfection and use within itself. Love is useful as well because love is a universal solution of many problems.

Special cases of perfection include: beauty, uniqueness, originality. Perfection is possible in four aspects: Physical perfection, Intellective power, Moral power, Power of money.

Special cases of energy: money, good luck, miracle, vital power.

8.2564. Functions

Ideas need energy, it's energy that puts ideas into material world from immaterial one.

Energy is a quality of thingness.
Energy is a quality of matter.
In order to make energy material, energy is needed.
In order to get access to its energy, an Idea needs a function-tool in the form of person-performer.

8.2565.

This world consists of energy, ideas and functions.

8.2568.

The more beautiful and useful an idea is, the more chances to find energy for being fulfilled it has.

8.2582. Single actions are nothing, a system and stability are necessary.

The system is very resilient, in order to make it swing it should be pulled with all might.
Do what you can... do not count on miracles or immediate results. Just do everything you can. The longer and stronger your efforts are, the more chances you will get.

8.2589.

For the system to work - one must work, the other must interfere.

8.2590.

One needs zero so that it does not relax and does not fall

into a self-destructive resonance. In fact, it is a fuse against the beginning of the chain reaction, turning into an explosion and self-destruction. The problem with thermonuclear fusion is that the reaction is uncontrollable, zero is what stabilizes the system, allowing you to keep control over the situation and avoid an explosion, while gaining access to energy.

8.2605.

The meaning of the harmony of zero and one is ten.

8.2634.

Unnecessary people pass by. Angels see only the light, the damned - only the dark.

8.2648. Similarities and differences.

One person differs from another one by a maximum of 5% or even less, the remaining 95% are all the same.

8.2741. Light is what is moving.

8.2747.

The same desires awake beasts in people. Competing for the same thing, they are ready to gnaw each other's throats. Civilization gave people the opportunity to wish different things.

8.2809.

Destruction is creating a free space for a new creation.

8.2849.

Sects do not do anything wrong, except that they immerse a person in infinite happiness, which turns him into a vegetable, and this is already drug addiction.

8.2853. The sound that you can touch.

Tangible sound... Square, triangular, round... sound folded to a point. It can even be licked, it can be as sweet as a gentle vanilla cream or as juicy as medium rare marbled beef steak. It can also be warm or excitingly cool...

All in all, you can do a lot with the sound if you can.

8.2855.

The Internet is not just a dump, it's a dump full of valuables. A talented scavenger can dig up a lot of useful things on the Internet.

8.2859.

Many people chronically think the same on completely different issues.

8.2876.

The person chewed in the TV is spatially unstable, but almost real already.

8.2895.

A true book writes itself. The books written by people are just miserable imitations of the true ones. The same goes for any other objects of art.

8.2903. Modeling the group behavior of nanorobots.

The person in the crowd models his behavior basing on the analysis of the behavior of the individuals near him. This is very similar to modeling the group behavior of nanorobots. We cannot control each nanorobot separately, besides, nanorobots are too primitive for independent thinking. Therefore, each nanorobot evaluates the actions of neighbouring individuals and does the same. And we control the whole herd at once through some simple signals, directing their movement and actions.

8.2905. Abyss of the feeble mind

Reading the Internet washes the brain out of the organism and leads feeble mind.

The reason is simple: the more primitive systems, located near the more perfect ones, degrade and computers are technically more perfect than people, so being around them, less perfect people get stupid. Communication with computer imposes certain obligations on a person, he just needs to be smarter than his computer, needs to work hard on himself, otherwise he will slide into the abyss of the feeble mind.

8.2907.

A lot can be said about a subject or a situation, not even analyzing it. Up to 90% of the information about the object properties can be found by studying what surrounds it, its origin, environment, analyzing the reactions of the objects surrounding it. In fact, to understand what it is, it's enough to look around.

External interactions determine up to 90% of the system internal properties. For example, gold-bearing lodes are recognized by the outputs of the accompanying minerals to the earth's surface.

8.2911. Cradle of filth and 800 MB of DNA.

People, just like computers ,can be paired and switched with each other, human even has special plugs, cables and slots through which information can be send.

8.2915.

The solution to the problem of cheap energy would allow not to dig tunnels under the ground, but to burn them out. It would also make it possible to create building blocks by simply cutting them out of stone. Remember, there are such things in Peru already, so, maybe, all this has already been?

8.2917. Scourge of God.

Mankind is the cause of many species mass extinction.

I've just thought. There are many versions of the extinction of dinosaurs and another 90% of species in the Cretaceous-Paleogene period 65 million years ago. Let us suppose another one: there appeared some kind of intelligent life, which eventually killed everyone. And an important consequence of the intelligent life's influence on evolution is a decrease in the size of animals. People try to kill large animals first and small ones are more likely to go unnoticed and survive.

8.2918. Thousands of masks.

Form and content... They are in balance, but their balance is quite unusual. At first glance, it seems that the form is much smaller than the essence... Moreover, the form is, in fact, a mask, a thin layer of illusion, hiding under a thick layer of real essence. How does this weightless mask maintain balance with the huge mass of its content? The fact is that the content can have thousands of masks. One and the same essence can have many masks, can take any form. The mask of the form is a matrix of thousands of variants, it takes by number, but there is only one essence.

8.2924.

Homosexuality is a natural phenomenon, connected with overpopulation and socio-economic relations development.

Non-standard oriented people reproduce less, consume more, work better and are rather efficient in the sphere of art. Reproduction and offspring consume a huge amount of energy that can be spent on other tasks, for example, spend more on creation-linked processes.

8.2935.

I see what I see. What I see is my subjective reality. However, you should always beat in mind that objective reality can easily have nothing to do with its subjective reflection.

8.2936.

You can see only your own illusions. The reality that lies beneath them can be anything, it should be taken into account before you believe yourself.

8.2938.

People are very similar to their masks. Masks are very similar to people. I would venture to assume that actually there are no people, masks are the people.

8.2941.

Any thing is highly dependent on the observer. In fact, 100% of the object properties depend on the observer's ability to see them.

8.2943.

You'll either be a slave of your goals or of some other's. The hierarchy of goals is such that big Goals subdue small goals.

8.2944. Preserved time.

We've learned how to preserve light, hiding it in electricity. Now I'd like to preserve time in Coca-Cola cans.

8.2947.

Basically, the issue of a man's love for a woman is the memory of the past life, he loves his past incarnation. A woman's love for a man is of the same nature. Probably, it's an issue of the incarnation procedure violation - this person was given the wrong body.

8.2948.

The future can easily create itself if it finds its Creator.

8.2964. Matter of mood.

We can assume that God and the devil are the same thing. Good God and Evil Devil are just the Creator's mood. While God is creating, he has a good mood, when everything has already been created and there is nothing more to create, God is starting to get bored and suffer... Anger and pain overwhelm him, he urgently needs to destroy everything to clear the place for the new.

8.2984.

It is better not to know about most things, it is enough to have a hunch of them.

8.2985. The characters spied upon.

People tend to parody everything. That's why there are so many fictional characters in our world, and many animals and plants are very similar to people.

All these trees, burros, woodpeckers, pigs, hens, horses, donkeys, wolves, foxes, lions and even insects are surprisingly similar to some people. How do you think, who of them is trying to be similar to who?

8.2986. The product of high technologies.

You are told that you are special, but that's not exactly true. In fact, initially we are all approximately the same, but you can really become special if you devote at least one of your life to it. Being special is not a gift, it is a very high-tech product, the result of a long and painstaking work.

8.2989.

If you don't like your life, just change it, you can do anything as you're God.

8.3025.

Human is an autonomous chemical bioreactor combined with a biocomputer and a number of peripheral devices.

8.3038.

The issue of homosexuality is an issue of memory of your past incarnation.

8.3039.

Within the justice and balance concept, I would assume that reincarnating in each new life, a person changes his sex of his body every time. In one life he's a man, in the next-a woman, then everything repeats.

8.3056. The 7th law of Soloinc. / About absence of competition for unique desires./

The more stupid and strange your desire is, the more chances to fulfil it you get.

8.3062.

As long as you believe in stupid things - you have a chance. Your stupidity is your air that keeps you afloat. Once you lose stupidity, you will sink.

8.3095.

Everything ends up with food... If everyone had enough food, no one would do anything... And we would quickly run out of food...

8.3113. In matters of the nature of things violation, the nature cannot be trusted.

The matters of survival of the weak and the health maintenance of the useless and the old contradict the ideas of evolution and renewal, therefore in these matters the nature is not an assistant to human, but his implacable enemy.

On the other hand, the nature cultivates strong, useful and beautiful individuals... Be in trend and the nature will help you.

8.3128. CD-Rom

Fun fact. Standard CD is 800 MB of information

During a sexual act, female organism through sperm gets 800 MB of useful information, necessary for conception of a child, which after uniting with female DNA in the egg, will form nearly 1.6 GB of information that will become the basis of the child DNA (two-layer CDs were of 1.6 ГБ, right for the human DNA to fit).

8.3129. Poet is dull without vanity...

Poets are the vainest people...
They are disliked as people, but loved as poets for that...

8.3130.

Many things would be unnecessary if they weren't necessary.

8.3150.

Probably, people are only required to create computers and provide them with energy. People are an intermediate form of life, created to maintain computers.

8.3151.

Getting into informational field, fresh and original thoughts become the heritage of other people.

That is, big human communities form some kind of informational field, which, among other things, influences brain activity of the people within it, especially of children and adolescents. The higher the quality of informational field is, the cleverer children will become. And the paradise is bad place for children to grow, there are too few thoughts.

Good informational fields are in big cities, science and art cities, in some intellectual or working communities. Atmosphere of the place has a great influence on informational field. Getting into informational field, laziness, despondency and uselessness poison it.

8.3152.

While travelling, human spreads his thought DNA over information field. Visiting new area, person connects to its information field, uploading there his information and partly downloading what there is. The more informationally useful and more perfect an individual is, the more privileged travelling conditions are created for him.

It's useful to invite outstanding people, original thinkers to your place. The very things they say are not so important, it's enough just to put them into your informational field to read a mass of interesting information from them. However, the opposite is also true. Connecting to the information field, a perfect observer is able to get everything there is out of it, synthesizing new information on its basis.

8.3153.

There is a range of groups of people, for example, children, adolescents, scientists, businessmen and artists who it's very beneficial to travel for. And there is a range of people, who nature creates favorable travelling conditions for: these are unique and perfect people. Inviting such people is very good for enriching your informational field. Basically, these people already put their thought DNA into environment just passing by. The challenge remains for unique and perfect observers able to read the information to be at the same place.

There are the following nuances. Informational field is a huge database, skills are required to find something useful in it, it's a long and painstaking process. Nature regulates this process initially, providing preferential treatment to processes, associated with information fusion. In general, standing near an outstanding person is good. And life makes outstanding people travel the world..

But the key here is to respect balance and measure, as information fusion isn't the very purpose, the very purpose is synthesis of a new thought or transformation of a thought into an action, thought materialization.

8.3154.

Humanity is basically a distributed cloud-based informational system with deep data backup and multistream parallelizing of information flows.

8.3155.

Nature provides preferential treatment for travelling to unique and perfect observers. It' linked with the system's need in information fusion, synthesis of new information and information materialization. The people who implement these function get preferential treatment for travelling, moreover this process doesn't depend on them much and goes some strange random way. For example, having made myriad travels within the Earth, I find it difficult to say how, why and what money it happened with. That's how my life went, without any special bases and reasons for it. Probably, the providence decided that being a bee, flying and pollinating this planet by thought will be helpful for me and created conditions for this act. Every city or country that I visit, generate a thought in me and I apparently leave part of my thought to them. Millions of thoughts were there in my head for many years, and years after I see their implementation all over the world. Evidently, having got into informational field, they were read from there by local perfect observers and implemented.

8.3156.

I endlessly walk the world and thoughts pop up in my head-here it should be done this way and there it would be good to do some other way. Some years pass and locals implement these thoughts ... Which is undoubtedly good and useful.

Providence, using me as a bee, makes me move from city to city, walk them for tens of hour and go to the next city at once. The need to move overwhelms me, it's almost impossible to stand still. I don't even know how it goes, where I get time and energy for it. For decades, the issues of my energy supply are solved miraculously without my participation, all that nature requires from me is endless movement and endless thought synthesis.

8.3157. Cloud anthill.

I wouldn't consider human as a self-contained system, it's rather a distributed cloud information system. Kind of a perfect beehive or anthill with various individuals that are basically semi-autonomous peripheries in the unified informational and energetic field.

Functions of the peripheries are the following: energy generation (sources of energy), population reproduction, protection and security, service and management, working and construction, new information synthesis, information fusion (pollination), reading information from the field and turning it into matter.

Basically, humanity is the development of the hives and anthills idea. There are sure to be many other roles in the anthill in addition to the roles of warrior ants, workers and those in charge of reproduction: thinkers, hunters, trackers, engineers. And anthill is sure to be a unified cloud-based system, too, less freedom and autonomy of separate individuals ultimately distinguishes it from a human. Humanity, unlike anthill, is more open and universal system with more freedom of separate individuals.

8.3158.

Informational field is heterogeneous and fragmented, it is divided into sectors. Information is kept in segments. Especially important are historical, scientific and spiritual places, where historically valuable people have been, each of them left part of his thoughts and soul on his way. Even more information is left in the places, where these great people died.

8.3159. Books can be burnt, but thought will live forever.

Once created, thought stays in the informational field forever and it will always get into people's thoughts from there.

8.3160.

Angels must encourage the good, demons must eliminate the bad.

8.3161.

Any angel is a demon at the same time. If a person sees something right where there is use, perfection, beauty, uniqueness or originality, he must support it, if there is none, he must destroy it. There are also 20% of individuals with an inverted value system, who work an opposite way. It's necessary to avoid fatal evaluation errors. The majority is usually right, but if it makes a sudden mistake, there won't be fatal information destruction.

8.3162. Useful thoughts are beautiful.

8.3163. Thought generates thought.

8.3175.

Reasoning is creating a thought. Thought is information that has characteristics of novelty for this organism. Reasoning is the process of information synthesis when new information based on the old one is generated. If this information has characteristics of novelty, uniqueness and usefulness, such reasoning is basically the purpose of existence of the information system and of human as its particular case.

8.3176.

Reasoning is a creative process and thinking is a mechanical one.

Reasoning is like riding and performing a duty and thinking is like idling. Reasoning is creating a thought and thinking is brain idling when a thought isn't created, but something old is moving in a circle.

8.3214.

The power of thought is such that it was used to create the Universe.

8.3223.

Many businesses are based on solving artificial, previously created problems.

8.3224.

Man is actually a modernized ant.

8.3226. Bamboo human.

I've noted that sects, plunging a person into endless happiness, turn him into a vegetable. The funniest thing is that this vegetable addict can't eat meat any longer....

By the way, speaking about meat, the matter is that without meat organism turns into vegetable even quicker. Meat makes human a predator and vegetables graze and are plunged in blissful idiocy and denying reality. Basically, they are not even vegetables, they are trees. Sects turn people into happy trees....

8.3227. Made-up person.

Before becoming yourself, you should make yourself up. And it's not as easy as you think, because you should become unique. It's very hard to avoid repetition in the world, where there are 8 billion people.

8.3229. Sex in the museum.

My walks in museums are like sex - mindsex. Thought DNA, crossing over, give birth to great new ideas. However it should be remembered that sex can be good and bad, and good sex is often associated with the unexpectedness and surprise...
*

There is a mass of things that have been preserved under a thick layer of garbage. It's worthwhile to dig them out and exhibit them in a museum for mindsex. Basically, the museum is a place for sex. The place where pleasure lives.

8.3232. Self-organized system.

God is the first slave of the world he created...

It can't be claimed that God rules the world that he created. The world is a self-manged system. Moreover, it even finds its creator and turns him into its slave.

The world endlessly builds and builds itself by its creators that it creates itself.

8.3234.

The Universe is billions and billions of small boxes, nested one inside another.

8.3236.

It can be assumed that ants are kind of perfect utopian civilization. Their elites live in the virtual worlds and slaves and warriors, devolved to the most primitive mechanisms, get energy in the real world.

8.3237. 20% of the Earth biomass.

By the way, there are much more ants than people on the Earth.

8.3239.

Anthill is the utopia, where elites live in the virtual world and slaves and warriors devolved to biomachines. Ants are very similar to nanobots.

8.3240.

Ant civilization has existed for thousands of years. Probably, real gods of this world are ant lords. Moreover, it can be assumed that humanity project was invented by ant elites on their model, as basically human is just a more perfect ant.

8.3247.

You should believe that there is no death. It will make your life easier in many ways.

8.3253.

When we speak about essence and thousands of its masks, an idea comes to our mind: can't it be so that there are only two variants of essence: zero and one, and other billions of its forms are just masks.

8.3257.

The individuals living in the virtual world don't need eyes

anymore, they don't need ears, they don't need anything but biomass, cooked by caring ants for them...

8.3258.

They are plugged in directly; they don't need TV anymore to look at the wall. Moreover,they don't even need eyes.

8.3261.

A musician is basically a chemist creating a drug, if the product is good, it will be a great source of joy and pleasure that will go like hotcakes and they will come asking for more and more...

And if the product is bad and it brings tiny bits of pleasure, nobody needs it.. However, there is a nuance. Drug dealers can't stand rivals; they will drag your item trough the mud and won't let it to their territory.... A war will break out...

Get it? It's a big business, one needs balls of steel and big guns to get into it.

8.3262. Burnout creators.

One shouldn't take offence at artists, their main virtue is their children - their artwork, the very thing they should be loved for. Bearing in mind that as people they have burnt out long time ago, giving their fire to their works.

8.3282.

Potentially and taking into account past lives experience, every person is endlessly old and wise since his birth.

8.3283. Childthoughts.

One can love the soul and even have sex with the soul.. I call it mindsex. Thought DNA of these people cross over and new fresh thoughts are born – childthoughts.

8.3310. Four types of information systems.

Biocomputers: trees and silicon computers: sand.

Deserts versus forests. Who will win ?
Maybe, the ocean?

In essence, we see two types of information systems.
Biocomputers: human brain, trees and so on..
And silicon-based systems: computers, sand, stones...

There is also a system that combines both of these technologies: the Ocean.

The ocean dominates this planet...
Humanity will be the fourth power in this case...

8.3311.

Wolves are good predators. Primarily, they eat the weak, the ill and the foolish. So if they want to eat you, there are good reasons for it...

8.3321.

Technologies progress much faster than people, and that's a problem as against the background of perfection, less perfect systems devolve and die out.

8.3322. Billion-core processor.

The more distributed and parallelized the system is, the less energy it consumes and the more vivacious it is. Humanity is basically a billion-core processor, unified cloud information system.

It's occurred to me - who said that 95% of human brain isn't used? Maybe it's used, just not by the human himself..

8.3328. Is speed of time a constant?

The speed of time is stable and depends on the speed of light. The speed of time can be changed if the speed of light is changed, the question is why…

P.S The speed of light is a variable…

8.3330.

Female brain is a multicore device where all calculations are chaotic and not dependent on each other.

8.3333. The key is not to overdo it.

Labor created human from a monkey. But labor also created some insects. For example, ants sleep 4 hours a day by fits and starts. Bees are buzzing constantly, too.

8.3334. Magic math.

Ten is divided by seven and three the best.

8.3377. Starting price.

How much is a thought? I sell it for thousand dollars apiece.

8.3381. Awakening.

A human personality is very scalable. One day it's dormant, another day a point generates the whole Universe.

8.3387. Warnings of an approaching apocalypse.

The following usually leads to the apocalypse and destruction of civilization: virtual reality, super life extension, fusion of man and the machine, super-humans and titans, self-learning artificial intellect, neural networks, crisis of overproduction, overpopulation, depletion of resources, merger of big, medium and small business and, most likely, thermonuclear fusion (one mistake is enough for another Sun to appear on Earth).

8.3390. Soon they will intent the way to swap bodies and there it can be left.

8.3396.

I wanted to tell the same in thousands of words. Yes, it's right, Dream rules this world, its best servants are Faith, Hope and Love. Their main weapon is Luck.

8.3405. Philosopher's stone.

Work is remedy.
Work is rest.
Work is pleasure.
Work is love.
Work is life.
Work is everything.

8.3414. Thought is the greatest heritage of humanity.

Human thought is the greatest achievement of evolution.

8.3415.

Power of thought is the greatest power in this Universe.

8.3419.

Gaseous metal is a distant relative of liquid metal.

8.3420. Antimatter.

The matter inert to time, existing out of time. The matter out of space and time.

Now they are trying to find an island of stability at the level of transuranic elements 113-120 in the periodic table, trying to go higher, increasing the core mass. But I think there still is the possibility to step aside, kind of slide aside, for example, new type of hydrogen - the one that could be in the other, parallel or negative Universe.

8.3422.

This world is nothing but construction, each and everyone builds his own and someone else's dreams here.

8.3423.

In fact, the person himself is an element of the neural network. If a group of compatible people is united into one team, they form a neural network. But I noticed that neural networks can also be formed with dead souls.

8.3430.

Insects don't feel pain, mindless fishes too. Beware of those who promise to kill all your pain.

8.3441.

Life is an exchange of the things you have in abundance for the things you have in scarce.

8.3442.

Hope is a good source of nuclear energy.

8.3448.

Faith, Hope and Charity give birth to good Luck.

8.3449. An inverted pyramid.

From a sacral point of view, a Goal (Dream) is a goddess whom Faith, Hope and Charity serve... Good luck, money, energy, happiness, health, joy, success... all of them result from the union of this three.

8.3462.

The essence of many series is that they are great to fall asleep under.

8.3483.

Feelings are chemicals and your brain is a chemical reactor.

8.3484.

Nothing disappears, but much becomes invisible.

8.3502. Wandering spirit.

I think that genius is the wandering spirit...
Sometimes it embodies crazy fanatics appealing to it and creates masterpieces by their hands... Human can't be genius, he can only be a prophet of genius, apostle of genius, slave of genius...

8.3523.

It doesn't matter what your God's name is, God has many masks, but only one essence.

8.3534. Person-shell.

Meaningless people can rarely see sense in something.
Everything seems shallow to a shallow person...
When I look at the egg, I see a bird, see wings and see clouds...

And what do you see? An eggshell? Scrambled eggs? Do you feel tender taste of breakfast and hunger?
Tell me, can you feel anything but hunger?

Emptiness demands filling, doesn't it?
Nice growling of belly and scrambled eggs for breakfast. This bird won't fly anywhere anymore. And we don't even know for sure if it could ever fly.

8.3559. A secret of youth.

A woman's youth and beauty are kept by the perspectives of reproduction. If a woman's partner is perfect from the reproduction perspective, then nature will take care of this woman in hope to get as many children from her as possible. Such woman will keep her beauty, youth and health for much longer time than it's usual.

A few more things that help to keep beauty and youth,

include healthy food, coldness, harmonious contrasting alternation of emotions like positive/negative ones, beautiful goals in life, interesting work, beautiful clothes and things around, physical activity, cosmetic medicine, endocrinology.

8.3560. A person covered with butter.

The Universe is egoistic as it does only what it likes and wants to, demanding people to serve its desires. But it likes loyal slaves who are ready to fulfil any of its desires. When the Universe loves you, it's a big happiness as this way good luck follows you. Health, youth, beauty, money, joy, happiness... if you behave yourself, they can even cover you with butter...

8.3563.

In "Variothoughts" it is written about trends, and those who try to argue with it, talk about exceptions.

8.3564. Dream conductor.

It's very helpful to be a conductor of other people's desires and dreams. When their desires are fulfilled through you, you make all the power of their energy work for you.

8.3575.

Genius is born contrary to and due to the resistance force. Reaching critical mass, resistance generates the supernova explosion.

8.3579. Genius is the person who managed to get out of his point.

Any person can get out of his point and become the Universe.

8.3581. Is child a genius?

The problem of creating a genius straight out of a child is

that Genius is the fanatic in love with his dream. Faith, Hope and love, combining in his heart, turn into the Sun. It's hard to imagine a child guided by Faith and Love. And a child-fanatic is a pure nonsense.

8.3588.

The more differences there are between the DNA of a woman and that of her partner,the better. The DNA of her partner gets into her body to keep her youth and health. Sex with such a partner is nice, his smell is arousing and children from such partner will be more genetically perspective and healthy. It's possible to recognize such partner by his smell.

8.3590. Good software.

In a person, like in a computer, everything depends on the software. He can be empty, he can be filled with cartoons and photos, but he can also design spaceships.

8.3592. The 8th law of SoloINC.

Whenever you may go, you will get where you want to. Any path leads to any goal. Just imagine your dream and go towards it.

8.3595. God's wish.

Kill the God if you meet him. Basically, death is the only wish of immortal God. Death is the only thing that he wants and that he doesn't have.

8.3608. A neural person with a filter in input.

We upload a data store to a neural network to teach it something. The human brain functions in a similar way: you have to upload something to it, say, a couple of books... There is one major distinction, though: the human brain perceives information in a distorted and arbitrary manner. The outcome of its work is unpredictable because the uploaded source data is entirely arbitrary. In other words, the human brain is a system with filters in input. Human brains are more or less the same, but they always deal with totally different source data.

8.3609.

The books written by people are soulless and thus, bogus... When a book has a soul, it creates itself on its own...without letting the author get into the plot and change the life there.

8.3621.

Light is stronger than darkness. Darkness always runs away from light. But light needs food when there's no food anymore, light dies and darkness comes again.

8.3622.

I've noticed great heterogeneity of time: it flies extremely fast on some issues and almost freezes on the others.

8.3645.

Basically, light is a source of energy, it's a nuclear reactor, it's a star. But the reactor needs fuel, moreover, the reactor shouldn't be stopped, as later it will be very hard to restart it.

Darkness is the food of light, it has a lot of fuel for energy-generating reactors. Light devours darkness searching for energy for its reactor

8.3650.

A person, as any neural network, is exactly what is uploaded into him. He is books that he heads, TV that he watches... people who surround him, but people full of social networks confuse me most.

8.3652.

A person usually believes only in what he likes, but he does not believe in what he does not like.

8.3655.

Talent is a matter of training and genius is a result of fanatic love and belief in your cause. Hope in this case is similar to foolishness and allows to ignore voices of reasonable people, daily claiming futility of foolish efforts.

8.3661.

Wherever you go, you will inevitably come where necessary, as you are destined to be exactly where you are.

8.3662.

The Universe asks for beauty.
The Universe begs on its knees...
The Universe draws its hands to you in hope...
And screams ...howls ...demands grinding its teeth with rage...
For you to make a squirm, go all out, but get more...

Beauty...

8.3668.

I'm not against eternal life itself, but for God's sake, can body be changed at least once in 70 years?

8.3669.

There should be nothing unnecessary in surrealism. Every booger in a painting should have deep philosophical message. If a booger has no message, something indecent should be added to it.

8.3671.

Beauty demands time. Beauty demands money.

Beauty is an amplifier of power, after swallowing energy beauty produces thrice as much of it. Having spent one dollar on beauty, you will get three of them.

8.3675.

Philosophy is a science about form and content of verity.

8.3678.

Being a primitive human for a while is a great pleasure. Civilization enslaves a person and deprives him of freedom, thus causing his pain. Taking off the chains of civilization, getting back to the primitive state for a while, blending into nature - all that is a great pleasure and rest of the soul.

8.3679. Freedom is a pleasure.

Civilization is pain. Civilization deprives man of his freedom... Sometimes it's good to reset the chains... and immerse yourself in a primitive wild state, full of wild fun and freedom.

8.3681.

Sex in the Russian smoke banya is a cosmic pleasure for a woman, opening all of her sexual chakras. In this way, she can even treat infertility. The spirit of fire literally permeates the body with ambrosial pleasure.

8.3691.

The issue of traffic jams in big cities can be easily addressed by building a second level of streets over the first one.

8.3701.

When starting to work on a highly important task in order of priority, the central processor refuses point-blank to operate with other processors, even the vital ones. In practice, it looks like a person absorbed in his problems starts ignoring altogether everything around him, including everyday issues. There's simply no room for this in his head.

8.3702.

Succeeding in the great, they devolve in the little.

8.3722.

People are pack animals and go in mobs. So any individual choosing «his own way» will undergo severe public criticism, especially one of relatives and acquaintances.

8.3728.

Everybody can be wrong. Listen, absolutely everyone can be wrong at the same time.

8.3729.

I'll clear up the point that everybody can be wrong. The key is that such a phenomenon as «everybody» basically doesn't exist. «Everybody» is an illusion. There are only some reputable opinion leaders that propose and create ideas and other people just decide whether they like them or not. ... If an opinion leader creates something interesting, liked by everyone, then this point of view is spreading around in people's heads. And this EVERYBODY, who kind of thinks so, is just nothing more than a group of people that likes something... This situation has almost nothing to do with the truth or a lie.

8.3735.

Faces are like rivers, going with the flow of life. And you won't step into the same skin twice. Every face of yours is new... Of course, you will be missing ...but only the future.

8.3759.

It can be assumed that human body is some kind of a bioprosthesis. Eternal gods have lost their own bodies long ago and like playing people very much, embodying from time to time.

8.3767.

People disturb each other very much. Human is the only herd animal that constantly needs solitude. Nevertheless, he can't be alone for long time. It makes him very similar to a dolphin that lives underwater but comes to the surface to breathe. Solitude is much like air, you gasp without it.

8.3768.

A lie is stronger than the truth, but a lie is afraid of time and dies quickly, while the truth is not afraid of time and can live happily ever after.

8.3770.

Genius is a wandering spirit, relax and let it possess your body.

8.3775.

Creator needs a muse. Muse is love. Love opens your eyes to beauty. It's rather hard to create beauty until you see and feel it.

8.3779.

I wonder who was God's muse while he was creating the world?

8.3796.

Do you think devil is the one with tail and horns? No, that's just a goat. Real devil is hard to recognize, it's the most powerful angel, he is endlessly diverse and his name is Legion.

8.3802.

The key is to work. You are a tool, you were created to fulfill your purpose and to work. If you don't work, you will howl with pain every second...

8.3804.

If a person doesn't seem like a monster to you, you just don't know him well. This person doesn't trust you and he never took off his mask in front of you.

8.3805.

They say fortune favors fools. On the one hand, it's a misconception, but on the other hand, fools are fearless. Fearlessness makes them heroes. Further there are two options: either he will die infamously or he is bound to glory and greatness.

8.3814.

The system is highly foolproof; it is as hard to break something as to change something for the better.

8.3860. Censorship of thoughts.

Society has approached a new stage of control over personality – censorship of thoughts. The authorities closely follow what people have inside their heads and suppress any thought that is at odds with themselves.

8.3869.

God can't stand when somebody gets offended with him or dislikes the world he has created. Think about it before expressing your discontent with something.

8.3873.

Artists are touchy, just like God, they can't stand criticism towards the world they've created.

8.3889. Devolution.

Monkeys don't want to turn into human as they are too lazy to work. If a monkey started to work, it would turn into human at once and the opposite is respectively true.

8.3892.

Evolution is a complex step-by-step process. Upgrading skills one by one tediously for a long time, it's possible to gradually become human.

8.3893. World of primitive gods...

It's occurred to me that life in the virtual worlds doesn't' include work. It's known that without work human devolves back to animal. Could it happen that virtual worlds are full of wild superpeople, devolved back into monkeys.

8.3899. Superdifficult task.

As for life goals, you should clearly understand that the goal that is much bigger than you, you will be crushed by it like the dog that the elephant sat on. Imagine that you, Zero and Noname, wish to win the hand of the Beauty Queen. On the way to this goal you will have to rise from Zero to Hero.

8.3901. Play but not too long.

Junk science is not science but a fun thing to do before sleeping. It's entertaining reading that brings joy and sometimes inspiration. In fact, it's a variety of science fiction, something that could have happened, but didn't. Nevertheless, rich imagination is a good thing. Lies and illusions is the basis of human reason; animals are incapable of that.

8.3904.

A talented person can spoil whatever he wants. The fact that the system is foolproof won't help in this case.

8.3907.

As long as a person just dreams, it resembles the Universe, folded into a point. But if you start to realize your dreams, it will be similar to the way God created this world.

8.3910. Everything is possible, the question is what for?

8.3911. Tuning beauty from the inside..

Clever facial expression is substantively more beautiful than the silly one. This conclusion shows that beauty is accessible for everyone and is not so dependent on the very appearance. The external beauty level can be tuned by the power of mind.

8.3912. Something to think about.

Work allows human to evolve intensively, while its absence allows him to devolve intensively.

8.3914. Humans are the only imperfect animals.

Humans are imperfect. Their own imperfection causes them great pain and humans become much better and more perfect by running away from this pain. Humans are the only animals that feel the need for perfection.

8.3929. Nevertheless.

Everything can easily be not what it seems. But it can just as easily be what it seems.

8.3931.

A woman is an energy storage that supplies those she loves.

8.3932.

A beautiful woman can be compared to a flower that grows in the sun and consumes its energy and later gives it to the ones she loves.

8.3933.

Any person is the whole world with seven billion of bacterium and a lot of other population living in it.

8.3934.

There is the only cause of everything and that is you. Probably, you are God, as everything depends only on you and on the strength of your desires.

8.3953.

Upgrading yourself is an expensive but extremely pleasant pleasure.

8.3954.

It's possible to endlessly extract some fresh thought from a good book.

8.3962. Good pragmatism.

Mysterious circumstances should be used to some advantage.

8.3970.

Time spoils plans, but not every of them. Such noble and exquisite plans as cognac, wine, jamon only get better over time...

8.3981.

Love is not only between boys and girls... There is also love for your cause, for your dream, love for life and so on...

8.3983. Beauty kills pain.

8.4001. Problems that make us stronger.

Do you ask Lord for power? He hears you and gives you

weights to train everyday and get stronger. When you gain strength, you will be able to achieve other of your desires.

8.4002. God hears you.

Look how it all works. When you want something but can't get or do it as you lack strength or abilities. The providence hears your desire and starts to help you.

The problem is that you're too weak for your desire, and thus you need to be trained. The providence starts to send you problems and troubles to overcome so that to make you stronger. As soon as you become strong enough, you will be able to fulfil your desire.

The fact that you lack money and not strength, doesn't change a thing. Money draws to those who have strength. When you upgrade your strength, money will get into your power. The main thing about it is to be persistent, resilient and not giving up...

8.4005.

It can be assumed that human isn't the end product of evolution at all, he is more likely to be some kind of error... System failure, some kind of virus that evolves and grows so fast that the immune system of the Earth hasn't reacted yet.

Probably, there is some virus which, getting into human brain, made it work some unprecedented way.

8.4007.

Mathematics is beautiful like life itself. The book of life is written in the language of mathematics.

8.4009.

The law is a matter of modesty: without laws, people will just grow more brazen and, fearing nothing, they will lose all sense of proportion.

8.4017. Beloved slave.

It is not quite correct to say "your Goals", it's better to say that goal doesn't belong to you, but you are one of its slaves, maybe even the beloved one.

8.4021.

I believe only in what could be true under certain circumstances.

8.4022.

Whatever God's name is, the key is that he exists.

8.4024. Civilization hasn't changed people

Given rather short time of human civilization existence and total length of evolution processes, it can be assumed that human hasn't changed over the last 10-15 thousand of years.

8.4025. Theory of evolution..

Not all people participate in the evolution, many skip it.

8.4026.

The essence of intellectual symbiosis between a man and a woman is that men are better at seeing the whole, while women notice more details.

8.4030. Crusaders outside the Byzantine walls.

Army of the allies going past your city walls is very dangerous, as they can change their mind.

8.4036.

Standard time of achieving excellence is seven years, sometimes if a person is outstandingly gifted, he can do it within three years.

8.4037. Laws of life.

God is law. God is some set of system rules and triggers which rule the world.

8.4042. Bad luck.

I've noticed that miracles not only help, but they may, on the contrary, disturb. The most certain plans are often miraculously destroyed. This kind of miracles is called bad luck.

8.4043.

Faith, Hope, Love for your goal ... generate miracles.

The person who doesn't feel all these three essences at the same time, won't achieve his goal...and miracles won't help him.

8.4044.

There are no constants ...there are only variables which can be changed for the good of the cause.

8.4045. In space.

Once, looking down, I realized that the Earth is very small.

8.4052.

It's hard to be a vegetarian in winter; bananas don't grow in the cold. It can be assumed that ice age made human a predator.

8.4061.

God, as any creator, extremely dislikes criticism towards the world he has created... If you don't like the World, you highly risk banishment of the providence...

8.4066.

Oooh …If only all the pain could be taken out of the head…

But it can't …as pain is human…

8.4074.

Contingencies that fill the life of a person say much about him…

8.4075.

It's quite common to remember past life…
But I've met those who remember future life…
Probably they just live vice versa?

8.4077.

Human shouldn't be blamed for being human. He is human, therefore he is vicious…

8.4078.

I woke up and my dreams, breaking the canvas of reality, woke up with me.

8.4082.

There is a very strict hierarchy in the flock of baboons. The closer society is to the primitive one, the stricter and more immutable its hierarchical laws are.

Civilization is more liberal in this regard. The less society worships its leaders and authorities, the more civilized it is.

8.4083.

It's immoral to punish a person for an attempt of a prison break.

8.4089.

Does person have a face? Person has only masks. One essence

and thousands of its possible forms.

8.4093.

You will be able to get where you want by any way, provided that you want it very much.

8.4094. Overload and emergency shutdown.

Something should be done with emotions, it's surely better to shut them down, but the system doesn't include this option, unfortunately.

8.4097.

Person's soul doesn't get old, but he marvels less over time and it ages him substantively.

8.4106.

People seem the same only from afar, in the vicinity they seem different... However, if you take a deeper look, they will become the same again...

8.4107. Ash and dust.

The idea that there was intelligent life on the Earth in the past is refuted on the grounds that no artifacts of those historical events have survived to this day. Researchers, however, overlook the fact that the nature of this intelligent life may have been different from human reason. As an example, its dimensions could have been minuscule... It could just as easily have been ants. Incidentally, reason could have been an advantage given only to elites, whereas simple insects could have been under control of more advanced species...

8.4118.

Joy is a source of pain. Love and happiness are joy. Having taken an ocean of joy and pleasure, you will know an ocean of pain and sufferings. Good things require a balance.

Of course, you can opt for calmness, but then you will not know love, happiness or other pleasures of life.

8.4121.

A 100% happy person turns into a brainless drug-dependent vegetable...

A person who is 100% sunk in quietude and deprived of joy, happiness, pleasures and compensating pain and suffering turns into an insect.

8.4122.

As a matter of fact, I'd compare insects to inanimate robots... These bio-machines have no emotions, no joy, no happiness, no pain and no suffering.

8.4124. Golden coin.

Running away from pain, you run away from happiness, too. Happiness and suffering are two sides of the same coin and they live only together.

8.4128.

The one who does good by one hand, does evil by the second at once. It's unknown if it's possible to do only good or only evil by both hands, but we haven't seen people with two right or two left hands....

8.4131. Times are changing, things are progressing.

There's an interesting moment, nowadays people are often kidnapped by aliens, but in the past they were mostly kidnapped by demons, devils or even angels...

8.4145.

One can't win over nature, but the very struggle invigorates and tones life some way.

8.4146.

There's a view that various parasites highly influence evolution, human in particular. All these mosquitoes and other things, by transferring foreign DNAs, including virus ones, from human to human (from animals to human.), highly influence his inner world...

8.4152.

The key feature of laws of nature is that they can be broken under no circumstances.

8.4159.

Artificial difficulties bring substantial diversity into grey monotonous life.

8.4161. Emotions producing factory.

Considering the World (within the concept of bored God), as a factory producing various emotions for the Immortals mired in eternal boredom, it can be assumed that specific people are specialized on specific emotions. The nuance is that your being specialized on production of pain and suffering will be very doubtful pleasure. So be very careful choosing the position of a permanent sufferer, having played this role well, you may further get only this kind of roles.

It's worthwhile to position yourself as a specialist in production of some more positive emotions, such as surprise, joy, pleasure, expression and so on. To do this, you should intensively cultivate the necessary emotions in your organism. If public and directors like the way an actor plays in his position, he will be given the same kind of roles further.

Anyway, an actor will be given the roles that he will play the best, according to the director. By the way, that's why, for the people who don't want to play dull roles, it's worthwhile to celebrate coming of great problems and difficulties by joy and pleasures. Firstly, it relieves stress and secondly, circumstances will think that there is no reason to set such people in dull movies.

8.4162. The power of love.

Any woman, once she becomes a Muse, is able to turn the man she loves into a Hero... Heroes are all mighty, they can control time, energy, money and good luck...

Heroes are the lords of the world and
Muses are the lords of the Heroes...

8.4163. The strategy of Impromptu.

Life demands chess strategy... while performing a study, at some critical moment it's necessary to step sideward and perform a sudden and beautiful spur-of-the-moment action.

8.4166.

Whether a person is good or bad highly depends on the distance between you... The person from afar and the person close by are two completely different persons.

8.4211. Note for a doctor.

People like to feel special, so the reference books on diseases are so thick. To be ill with something especial is somehow even more pleasant.

8.4233. Genius in love.

It's impossible to teach and raise a genius, as the real genius is a blind and deaf fanatic, endlessly in love with his ideas.

He simply doesn't hear, doesn't see and doesn't understand other people, doesn't get information and messages that are unnecessary for him. People see the majority of geniuses as extremely strange, often stupid people who devote their life to some nonsense and ignore any reasons and attempts to guide them in the path of righteousness.

8.4242.

The soul has no age. Souls are immortal and eternally young.

Age is something that has to do with the body, whereas mature souls are, basically, still the same age and differ only by the amount of their internal knowledge and experience.

8.4262.

Even the truth can't be trusted, because as soon as it loses its circumstances, it gets spoilt at once.

8.4274. Fictional character.

I don't know whether God exists, but I find existence of human highly questionable. I presume human is a fully fictional character.

8.4277.

I found a vacancy in the cinema: nobody has filmed aphorisms and essays yet.

8.4284. A highly-qualified garbage man.

One more subject should be introduced to the school curriculum - Introduction to Reading the Media.

The Internet, newspapers and television require the reader to have been trained to a special qualification – that of a garbage reader. A highly-qualified garbage man should be skilled at sorting out the following types of information: bullshit, advertising, manipulation, propaganda, taste modifiers, chemicals, sticky sweets, zombie garbage and other dirt.

8.4346. I don't like people, so I don't eat them.

8.4347. Cannibals love people the most.

8.4348.

The farther your nearest and dearest is, the easier it is to love him.

8.4366. Cause of humanity dying out

New generation of virtual men and women are much better than the real people. Virtual people can be superheroes and have any appearance, the real ones wouldn't be like them even in their wildest dreams.

8.4382. We need surplus for exchange.

Superfluous created civilization. Human civilization is the civilization of superfluous things. Surplus is necessary for exchange, it created trade. If everybody had only what is necessary, there would be nothing to exchange and civilization would die along with merchants, workers, peasants and even women would lose the reason to stand men.

If nobody can give anything to anyone, nobody needs anyone.

8.4387.

Many people live in their illusions and one shouldn't pull them out of there for nothing. A mollusk, pulled out of its shell, dies.

8.4420.

Evil can be beaten by love, as what we love can't be bad, what we love is always good.

8.4426.

People are ugly... from the point of view of any other reasonable creatures.

8.4428.

I have never met the very fabulous ordinary person yet. On the contrary, everybody I've met was overly special...

8.4456. However...

This world is ruled by men, and men are ruled by women, while women are ruled by their fitness or yoga instructors, cosmetologists, female books writers, internet-reads, tv, mothers and friends.
... I even start to feel somehow scared...

8.4463. Zombie-prone.

TV is a very dangerous thing, having it at home is highly zombie-prone...

8.4479. Deeply-rooted traditions.

Culture has only existed for ten thousand years, while inculture - for millions of years.

8.4485. Art needs master.

Art can do without a spectator, but it will endure hard times without a master.

8.4487.

Creativity is a disease that makes the ill create medicines for himself by his own hands.

8.4488.

The power of a work of art is defined by its ability to create its own audience.

8.4519.

A beast lives inside every person and fear is the only thing that keeps this beast obedient. If the beast wins over fear and breaks free, it's like gates of hell open and the evil enters our world. Therefore, fear is a weapon against the evil. The fear of punishment is the only thing that can stop the beast, living inside a person.

8.4533.

The duller the movie is the closer it is to reality, respectively, the more nonsense it has, the more interesting it is.

8.4541.

A conservative is the one who believes that a man should work, and a woman should be beautiful. Liberals are more sympathetic to sex minorities and insist that a woman should work .

8.4588. Radioactive genius.

Education is like teaching neural net or bombarding uranium with neutrons to create such heavy transuranic elements as plutonium and so on. The more knowledge a student will get, the more significant he will become.

It's also occurred to me that though super-heavy isotopes are likely to dissociate and don't exist long, nevertheless, a stable living genius can be created if you jump up to the stability island.

8.4589.

Genius is the star, inside which a nuclear reaction occurs. Genius is radioactive and dangerous when it's close.

8.4615.

Energy is only accessible to beauty... Beauty is the only source of energy... If you need energy, you should serve beauty...or you will die of hunger...

8.4623.

Life is pain. Do you want to rid me of pain? Do you want to kill me?

8.4641. God is vain... As only the vain can create...

Since God is vain, he likes being praised ...and likes beautiful temples...

8.4642.

Poets are vain, as only great vanity can challenge the titans of the past and the present. It's impossible to create without vanity. Vanity is confidence that you will do even better than the greatest...

8.4646. Creators are vain. Creator should be praised to create.

The same is in life ...If you want God to love you, praise him.

8.4650. Dream protects its lover.

Marshal's baton in the soldier's backpack is not only vanity that gives wings, but also a beautiful distant goal, that can save its owner from a stray bullet, as who will become a marshal if a soldier is killed at once.

8.4656.

An optimist is the person that hopes to benefit from every outcome.

8.4670.

This world is ruled by ideas and dreams...
Human beings are always slaves to their own or others' dreams.

8.4679.

Humankind's greatest blessing is his ability to forget and to die.

8.4698. Breaking through the closed cycle.

The paradox is the cycled essence when consequence becomes its own cause. A miracle is necessary to break the vicious cycle. Faith is the source of miracle ... love and stupidity are the sources of faith.

On the other hand, you can also die. Probably, breaking paradox is death. Probably, paradox is the very miracle generated by faith.

8.4699. The paradoxical Universe.

Paradox is a variant of perpetual motion machine. Basically, paradox is the basis of the Universe's existence. Paradox is the endless source of motion energy.

If you break paradox, you will break the Universe.

8.4700.

Paradox is a brood of faith, you believe in it, therefore it exists.

8.4701.

Can something don't believe in exist? Because if it exists, you will have to believe in it...

8.4705.

Do you resent double standards? Calm down, they don't exist... In fact, there are tens and even hundreds of these standards...

The more valuable a person is, the more he gets. The closer or more loved a person is, the more he is allowed to do. The more beautiful and excellent a person is, the more he is allowed to do... The level of allowance is function of a person's value level...

8.4713.

You can use paradoxes as a source of energy. Paradox is a perpetual motion machine, a source of endless motion.

8.4715.

The truth that doesn't match the circumstances is a lie.

8.4724.

Since writers have stopped writing books, they are read much less... Probably, it's due to the fact that writing was hard work and writer had to think over each word. And nowadays, when everything is computer-printed, there is no need to think, which has highly affected books quality...

8.4739.

Many things don't demand understanding as they have no sense.

8.4740.

Order aspires to self-destruction.
Order is what aspires to chaos.

8.4753.

Life is a dirty thing, while illusions are clean and beautiful.

8.4781.

People are the gods that went wild. Virus of Alzheimer's disease killed their memory and they forgot their past lives.

8.4782. The first human God created.

God created the world and started living in it under the guise of an ordinary man. Originally God created only one human - Eve. All the other people emerged by independent reproduction.

8.4783.

God is a man while a woman is the creature of God. Though, perhaps everything was the other way round.

Perhaps, it's not male God who created a woman, but a female Goddess created the world and then the man so that he would serve her...

Female Goddess came down to Earth and became the first human being and progenitress of mankind.

8.4784.

It's more logical to suppose that mankind took origin from a woman rather than from a man. Women are able to reproduce without men, at least by conceiving through the Holy Spirit.

8.4785.

A man is a simpler version of a woman that was created to work for a woman and her descendants.

8.4814.

It's more than likely that people took origin from a woman rather than from God. But perhaps, that woman was a female God.

8.4815.

Love is a source of energy.
The more people love you, the stronger you are.

8.4826. Freudian psychoanalysis.

The idea that parents and hard childhood are guilty of all your problems is very soothing and even inspiring.... The idea that all your problems can be pinned onto the others is very pleasant and joyful.

8.4833.

Humankind is an electric engine and need electricity to work. Desires are the main source of energy in the human soul. The power of the wish determines the current power and output.

8.4849. Hero is the killer of giants.

8.4850.

Imitation of thinking is favorite occupation of the people after imitation of working.

8.4854.

It's necessary to win time, time is money.

8.4857. Rain.

The future comes to the present by drops. Drop after drop becomes the rain. The rain, intensifying gradually, becomes a wall of water....

8.4858. Time is liquid... Time is water...

The present is an ocean with the past in its depth, while the future comes on the surface like the wall of rain. Drop after drop the future mixes with the present...

The future is a little higher than the line of the rain, the future is clouds. It's not physical yet, but soon it will turn into drops.

8.4860.

When something dies, it comes down to the bottom of time, plunging into the sand, turning into the sand....

8.4864.

I've understood why they make them live a long life: they want death to become joy for them.

8.4868.

People around you and your place of residence exert a tremendous impact on your thinking, because the human being is a neural information system that operates in semi-autonomous mode in cloud networking. It has nothing but what you've uploaded to it.

8.4869.

Time is uneven and heterogeneous. Time is a heterogeneous liquid, distributed unevenly. The natural circulation of time is stable, but uneven, somewhere it's raining, somewhere there is draught... somewhere time accelerates, and somewhere it slows down...

8.4870.

Given that time is water, 80% of human body is time... Dying, he can get to the clouds...

8.4872.

A beautiful theory resembles a woman, she's so attractive that any other of her faults are insignificant.

8.4874.

Consequences are breed of causes and circumstances. To change consequences without affecting their causes, one should change circumstances.

For example, night street lamps, cameras and police officers, as well as sport, education, work cause good influence on street crime.

8.4877. Hydraulic impact.

Time is liquid and it has all the qualities of liquid.

8.4884.

Man is public, not private, property. Private property in relation to man is abolished, but public property is maintained.

8.4910.

Peace is fiction, It was Einstein who proved that there is no absolute peace in this world.. Therefore, there is no death, even after dying you will keep moving....

...the parts of you will sprawl in the different directions and become part of a new life.

8.4914.

Everyone has1% of talent and there is no need in more...

8.4915.

I'm not the first philosopher in human history, but I' m undoubtedly the last.

8.4923.

It's true that our God is suffering... There are two reasons for it.

Firstly, he is immortal. Life is suffering and endless life is endless suffering.

Secondly, he is God-creator. Creating is a great source of happiness, but the greater your pleasure is, the greater will be the suffering compensating it.. So God often suffers... And his suffering is truly the pain of this world.

8.4926.

Humans must work to pay for their pleasures. Humans must work to pay for their vices... their happiness... their wishes...

I'd suggest that humans are means or tools that make their

vices, wishes and pleasures exist.

8.4938.

Human is God's toy, God likes going down to Earth to play people.

8.4955.

Although there are no miracles, they still happen periodically.

8.4967. A coin has three sides...

8.4972.

Whoever you ask, everybody considers himself to be normal, but highly doubts normality of others.

8.4974.

Equality is a terrible injustice.

8.4975.

All people are potentially equal, but not more. Somebody is still a point, while somebody else is already the whole Universe. The issue of equality is the issue of growth, somebody has already grown and somebody else is still growing.

8.5002. A counterintuitive source of energy called beauty.

The source of energy in this world is beauty. Beauty has the power to create energy out of nothing. Beauty is also connected with perfection, uniqueness, originality, use and even love...

If you need energy, money, good luck, health, happiness and other good things, then you should somehow serve beauty... you should love it and forgive it, work for it and share it with the whole world. Your task is to spread it.

The stronger the beauty that you serve is, the more energy

you will get from it. Beauty is a counterintuitive system, the more it gets the more it gives. Thus, the more they love it the more people serve it and the stronger and more energy-providing it gets. Your task is to make your beauty known and loved by as many people as possible... The sun should shine over the whole world. Beauty is a star, is the Sun... Your task is to make the Sun rise over the whole world.

8.5003.

If we take a look at the thought about beauty (perfection) as some paradoxical source of energy from the perspective of a relationship between a man and a woman, we'll see the following. A loved woman is always beautiful and perfect... Thus, it's very useful to serve her. The more a man loves or serves a loved or beautiful woman, the more energy she gives him. Beauty is in essence, God. It's necessary to worship beauty to make it stronger and more useful... beauty demands beautiful temples... Everyone should admire it. That's why a man should provide a loved woman with a beautiful house, car, clothes and entertainment... Should constantly provide her with traveling the whole world so that many people would see her and admire her. The more people admire her, the more energy a man will get. As beauty is very grateful to those who loves and serves it. Beauty is a great source of energy, happiness, good luck and pleasure.

An important moment- since happiness is a drug the dose of which should become heavier, you have to provide the constant growth and influence for the beauty you serve. Once you start it, you'll be unable to stop. If you betray your muse, you'll simply be damned.

8.5004. What can be a counterintuitive source of energy.

Beauty and perfection which possess the attributes of uniqueness, originality, use. Works of art can be beautiful (from music to sculptures and architecture), beautiful and loved people, including children, people interesting in their mind, looks, ideas, both men and women. Beauty can be an attribute of both men and women. Besides, mutual love gives birth to star systems and galaxies. Beauty is in ideas, thoughts, projects, things and people...

8.5005.

Love can create the Universe from a grain of sand. A handful of sand in caring hands is stars, constellations and even the whole galaxies.

8.5006.

Any man can by love, faith and hope create a star and even the Universe, becoming its God...

You can be God of many Universes and thousands stars. Wherever you see beauty, you should help it.

8.5007.

Lack of love and beauty is the cause of bad life and great suffering of a person. An endless source of energy is hidden in beauty and perfection and this energy can only be obtained by love and serving beauty. The closer beauty is, the more it is like the Sun and the more energy it gives. The light of distant stars is beautiful and useful, too, but weak... Any beauty demands love, but the closer should be loved the most.

8.5009.

Life is God's long sleep and death is his awakening.

8.5010.

The information field of the planet Earth resembles the Internet and the human mind a computer connected to the Net... Phrasing questions correctly and accurately is necessary in order to get information from the Net... The power of the wish, intense thinking and right questions generate solutions. Also, thoughts and souls don't die but are stored in the cloud.

8.5019.

The variety of religions is the issue of jealousy... They all love God, but don't love each other.

8.5050.

Should I prove anything to you? Tell me why.

8.5051.

The incredible comes true much more often than credible... It's amazing but true.

8.5063.

There is no such a thought that can't be misinterpreted.

8.5064. Hermit.

A musician (a poet, a painter, a writer) should create (create his work) secretly, reclusively and in private. This will let him grow and improve being happy and peaceful, getting energy from his works. You shouldn't show your works to anyone until you grow and gain strength, otherwise people's indifference and their angry criticism will get you into hellish pain and despondency, exhausting all your energy. And you won't be able to grow. And you will wither. And you will perish.

8.5065.

A person's price is just the sum this person manages to sell him-or-herself. Perhaps, one would cost more if paid more attention to advertisement and sales, but alas, laziness doesn't let one do it... Lazy people become sold at a discount.

8.5069. Dematerialization.

Being yourself is very dangerous when you are nobody.

8.5070.

Being yourself is prohibited in public places.

8.5082.

Life is the circle that never closes, as it's a spiral.

8.5110. Miraculously.

Majority of people don't need explanation of the way plasma engines based on thermonuclear fusion work, referring to miracle or science is enough. By the way, miracle and science are almost the same and magicians and scientists are interchangeable.

8.5125. Three types of reality.

It's possible to live in reality, in the fantasy world or in the digital reality.. Life can be real, illusory and digital... And we don't know whether they differ from each other.

8.5146.

Changing reality by power of thought is prohibited by the rules of this game, but sometimes, if one wants much, rules can be broken.

8.5171.

Illusions often hide the truth...
Illusions are the mask of the truth...
The truth is awful...
So it has to wear masks...

8.5223. God's philosophy.

Aliens' philosophy is unlikely to differ much from the human philosophy. At least Syntalism philosophy is great at understanding the nature of any phenomena - from sand and trees to ants and people.

8.5236.

Space is relative, so is time, the same goal can be reached by thousand steps as well as by one.

8.5253. Life is the fight for the energy.

8.5256.

Virtual love will obviously become the cause of humanity extinction. Virtual people are perfect, real people seem not so appropriate for sex after them.

8.5292.

I can explain to you whatever you want, I can answer any question. But it's not what you need. You don't need explanations; you need me to prove something to you... But you should prove everything yourself.

8.5294.

Once I died and, having opened my eyes in a new life, I realized I am the last person on the Earth, «Again... - I though -. Something should be changed»

8.5295.

Time is non-linear. Reviving from life to life, you will not necessarily live forward in time. Revivals are spontaneous. You can revive anytime, anyplace and as anyone. Moreover, various versions of you can exist at the same time and even be of the opposite sex. Probably, such a phenomena as finding your match, finding yourself are linked with this very phenomenon.

8.5297.

Given non-linearity of the time and God's ability to regenerate into any person, anytime and anyplace... Could it be that

each person is one of the God's revivals... All the people living on the Earth are the same individual.

8.5302.

You always move forward whatever your direction is. The
door behind you is not at all the door it was a moment ago.
Even stepping back, you will get in the middle of nowhere.

8.5303.

One can't step twice into the same river, as well as step out
of it, as the water has already gone by. You step back, trying
to go out through the door you came in, but there is no this
door, the other door is at its place.

8.5312.

It is rather easy to destroy the evil, you just should kill the
good.

8.5316. The more people, the more cannibals.

8.5319. Hyperlight speed.

Let's assume that the speed of light is finite and absolute,
but if the object that moves with the speed of light also
spins with the speed of light, would the multiplying effect of
these two speeds emerge?

8.5320. Era of the whores rise.

In ancient Rome whores were cheaper than now ...their price
was about one tenth of a specialist's daytime salary. Nowadays whores are unforgivably expensive... That is modernity can be called the era of the whores rise.

8.5324. Global warming.

The weather is closely linked with civilization, for example,
antic climatic optimum since 250 B.C till the fourth century A.D. ennobled the Roman civilization.... And tempera-

ture drop to the 5th century coincided with the elimination of the Roman Empire. Probably, a temperature drop has become the prime cause of the Migration time. Then there were long periods of decline, linked with the little Ice age. By the way, though global warming is now talked about, average temperature now is less than in the times of Ancient Rome.

8.5326. Religious pragmatism.

In the Ancient Rome water had to always be flowing on religious grounds... Water should flow for sanitary grounds, too, otherwise it spoils quickly...

8.5327. Sanitary pragmatism.

In ancient Rome community toilets were free for a long time as their owners collected piss and sold it to dyeing plants.

8.5337.

Over the last thousands of years human hasn't got cleverer at all... I can assume that there are even much more cretins than before at least because of the general increase in the number of people.

8.5341. Bosch's «Haywain»...

Hay... Cattle can take only hay from life ...

8.5375. Emoticon generation.

Fashion has changed. They used to wear masks, now they only wear emoticons ...

8.5402.

The key difference between life and game is that the game can't be stopped if you are tired or dislike something. That is, life is a game and many things in it are similar to a game but this game is very specific... often unpleasant , obscure, nasty ,obnoxious and painful ... So it's wrong to compare it with the games that bring entertainment and rest.

8.5403. The first after God...

Humans' desire to "do everything backwards" is a systemically important function sewn in their brains by nature. This function's essence is as follows: the main property of the mind is its ability to change rules. Animals are slaves to their instincts, whereas humans are capable of anything, if so desired... This makes humans similar to God.

8.5404.

Reason is what can break old rules and establish new ones.

8.5405.

Humans are robots that are capable of reprogramming themselves of their own free will and of breaking their own rules...

8.5406.

What sharply distinguishes humans from animals is that humans can break rules and animals can't.

8.5407.

The fact that it's hard to program humans 100% differentiates them advantageously from robots... Humans always have a chance to crush any program installed in them.

8.5469.

Decent people don't die, they only change their form.

8.5472. Principle of motion.

The key principle of motion is that it should never stop. This process is implemented by energy conversion from one type into another. Motion is the process of constant energy conversion from one type into another.

8.5499.

Yin and Yang are not the male and female essence, but rather the light and shadow that arise on the slopes of a mountain when the Sun rises and sets over it. Man is the boundary between light and shadow.

8.5500.

There is hardly a thing that creates more evil than the good.

8.5511.

The soul is software, the brain is hardware and the body is the peripheral control unit.

8.5515.

Lao Tsu is an old-newborn, an interesting image showing life as a transitory stage, life is preparation for life...

Wisdom of children and wisdom of the old. Some haven't put chains on their mind yet, others have already managed to get rid of them..

8.5525. Reasonable philosophy is the driving force of evolution.

They say evolution is a slow process, nevertheless I've seen the people that evolved from insects to rather sensible primates over 7-8 years just by embracing reasonable philosophy.

8.5526. Personal experiment.

To find 23 000 reasonable thoughts, I had to do the opposite 69 000 times and see ... where the wrong way leads.

8.5527.

There is such a phenomenon in poetry as "a library of

tropes". While "Variothoughts" is the library of the essence and forms in philosophy.

8.5529.

Was Prometheus the Devil fallen from the heaven ... bound to eternal suffering???

8.5530.

An energetically hungry person is stupid... Thinking consumes very much energy. Nevertheless, energy flows can be redistributed, cultivating some to the detriment of something other.

8.5531. The incredible variety of Beauty.

By cultivating imperfection, civilization contributes to the emergence of various types of perfection. In the past, natural selection used to cultivate only a few types of perfection and the rest perished... However, now that the weak have the opportunity to survive, the burning issue for it is the search for available niches to achieve perfection.

8.5532.

Imperfection hurts. A person who is imperfect in many things is sure to be perfect in something special.

8.5548.

The physical takes over the spiritual, as we live in the physical world, because physical bodies want to eat, dress and have a roof over their heads. Moreover, spiritual values, such as education and beauty, for example, are extremely expensive. And we remember that thought (spirit) aspires for materialization.

8.5553. Data carrier.

Human is a data carrier, uploading and downloading information brings him pleasure ... Therefore, human likes chatting, likes talking over and likes watching and listening to something interesting,. Processes associated with fulfilling of his purpose make human happier.

8.5564.

Death of civilization will be the cause of death of humanity. A highly specialized human is unable to survive out of civilization.

8.5566.

Sophistication of civilization leads to simplification of human. Universal human turns into highly specialized human.

8.5582. New world.

This world is based on the word of God. The technology of creating a new world by word is called literature.

8.5591.

The mixing of Homo sapiens and some Neanderthal is called «European type white man». Even more Neanderthal genes created Asian type man... Pure Homo sapiens stayed in Africa and moved nowhere.

Philosophically, this fact approves of travelling and mixed marriages.

8.5592. The essence of the human mind is stodginess.

A stodgy person is homo sapiens.

8.5679.

There are many bots in the Internet, this pseudomind is very similar to the real one, but it is hardly trustworthy...

8.5685.

The people who know life from movies are one evolutionary step higher than those who get knowledge from cartoons... But three steps lower than those living in the real world.

8.5690.

They say that there is either the truth or a lie... It's partly true, but, «the truth» is a matrix array, it has thousands of faces, as well as a lie. But the most paradoxical is that the truth and a lie are the same, the difference between them is associated solely with external circumstances...

8.5727. Hypothesis isn't a proof.

There is a theory that the Moon is as small as a pancake, and though at first sight it seems credible, there is no proof to it...

8.5753. Autumn doesn't prevent from being hungry.

There's an interesting moment. The apple that the devil fed to human was likely to be slightly rotten and already fermented. Basically, having started to eat rotten fruits off the ground (out of laziness, evidently) human discovered alcohol... Laziness is the cause of everything, human was too lazy to climb the tree for a good apple and he preferred to use freebie.

8.5754.

Rule is thing that repeats systematically, no matter how often, the key is systematically. An array (set) of rules can be attributed to the same system.

8.5764.

The right answer to the question «Who came first - a chicken or an egg?» is the following: Eggs came first, as God created Adam first.

8.5765. Superinsect.

Superman is the global embodiment of the idea of freebie. You sit doing nothing, a spider bites you in the ass all of a sudden...that's all ...now you are a supergiant insect ...

8.5769. Civilization generates garbage.

There is less and less of the valuable and more and more garbage.

8.5865. Death is a pause between two lives

8.5871.

It took only one moment to create the Universe. And it's being upgraded for 12 billion years already.

8.6032.

Human was not exactly exiled from the Eden, more likely, he, having fallen off a banana palm, couldn't climb back.

8.6033.

Reason is rather a by-product of pleasure-hunting... A candy is a good reason to reflect upon it.

8.6034. Training and philosophy are more important than genetics.

8.6036.

People are required to provide energy for computers and work for virtual game words and artificial intelligence systems that live in them.

8.6037.

Once a group of programmers created a virtual universe, inhabited with pseudoreasonable characters ..At first there was support service in this world, where one could send complaints and suggestions, but then locals were so annoying with their whining and spam that this option was switched off....

8.6038.

Once an emergency occurred in one of the virtual universes created by our group of programmers. Local pseudoreasonable characters, having decoded the system code, got out of their virtual world into global parental system and nearly took over our world. A lucky chance saved us... At a certain stage they managed to get to the higher level, where they faced our own creators, that already had the experience of fighting this phenomenon.. Launch of the system throwback allowed to manage the situation elegantly...

8.6056. The model of the future.

A dream is the creation of virtual model of the future aimed at modelling various situations and behavioral patterns in some particular conditions.

8.6106. Trees.

It will end in cultivating people like plants...

8.6107.

They are unlikely to reach the stars, they are more likely to perish in the virtual worlds, as they always want to choose an easy way.

8.6108.

A virtual world and a world of illusions are, basically, one and the same thing... It's just that the former exists in the computer and the latter in the human mind.

8.6110. There is less radiation underwater.

Basically, we all live underwater on the depth of 10 meters.

Air humidity is such that the atmosphere is nearly a 10-metre layer of water.

8.6111. Underwater world.

Basically, the Earth's atmosphere is highly discharged water in the gas cloud...

8.6120.

I wonder whether God is the only one or there is a race of gods. Are they all smart and perfect or somebody is smarter or sillier, more honest or more cunning?

And if he is only one left, isn't it a sign of degradation and imperfection, as the worst die out and the best survive and continue living.

8.6130.

All I can say on Egyptian pyramids: there is myth and deed in them, but no miracle.

8.6186. Choose where there's less competition.

The idea is that you can do this way and the opposite way, and both the variants will be right...

8.6188. Multimasked person.

They say people often wear a mask ... But it's more correct to say they wear masks.. One masks hides the second, the second hides the third...Masks hide masks.

8.6199.

Reaching enlightenment means learning to draw use even from useless things.

8.6208. Planet of illusions

Only God lives in the real world, all the others live in the world of his illusions.

8.6305.

If God is the only survivor out of the race of gods, it's obvious that this race reached a creative evolutionary deadlock and died out, which is undoubtedly not the best role model.

8.6333. Cultural layer.

Time is rather physical, for example, 2 thousand years are approximately a 2-3- meter layer of soil, those who attended a dig saw it.

8.6334.

A cultural layer basically consists of garbage, dust and remnants of the dead bodies.

8.6337.

People of the Neolith, the Ancient world or the Roman era were not sillier and maybe even smarter than modern people. Smarter, as they had to think much more intensively than now to survive... Now one can think nothing at all, lying on a sofa.

8.6344.

Egyptian pyramids were some kind of sacrifice to gods. It's very useful for rich people to build or invest into something epic and beautiful, it makes karma better and brings good luck. The Providence really likes when something amazing is created and it helps people who create the amazing...

8.6364. Programmed Food.

You shouldn't be guided nature in everything. Nature often needs food and you don't want to be the food at all. According to nature, you are the programmed food created to be eaten, . All your instincts call for you to be eaten. If you are eaten, you will be happy. So, think whether you need it or not...

8.6369.

The evolution of human civilization is the evolution of philosophy.

8.6385.

Even though civilization hasn't contributed to human evolution, it led to the evolution of philosophy. People haven't changed, but their philosophy has become more perfect and stronger.

8.6386.

Civilization is the evolution of thought, the evolution of philosophy.

8.6387.

The philosophy of a person is his operating system: the more perfect it is, the stronger the person.

8.6388.

Civilization hasn't resulted in a significant evolution of hardware, but it has caused a major evolution of software. That is, the human mind has remained almost unchanged, while his soul and philosophy (operating system) have significantly evolved.

8.6391. A simplified person

The development of human philosophy is comparable to the evolution of software: first, everything was getting more complicated, and then they have opted for interface simplification...

The Windows-oriented person is much simpler than the one who has worked with common line codes...

8.6392.

A Web 2.0 person is extremely simple and positive: he's got nothing redundant, but also very little useful stuff.

8.6393.

An object-oriented person is programmed much easier than a procedure person.

8.6395. I want to upgrade humans.

The goal of establishing the "Variothoughts" is to reprogram human philosophy. Quantum philosophy is a program package for upgrading the operating system, on the basis of which the human mind functions.

8.6397. Mind reboot

The ideas of quantum philosophy need to be disseminated and promoted. The more sound people live in society, the stronger humankind will become. Considering the challenges facing humankind in the third millennium, the survival of civilization depends on the good sense of each individual.

8.6398.

Miracles demand huge concentration of capital.

8.6548. The sense of life is Joy.

The sense of life is to have the right to enjoy life, doing your own job well. Joy is the payment for the job well-done.

8.6801. Creator either creates or self-destroys.

8.6808.

Creator self-destroys while creating, as it's necessary to put a piece of himself in each of his works.

8.6809.

God died after creating the world and human, but he stayed alive in his creations.

8.6821. Reptiloids.

Dinosaurs didn't die out, they become smaller and got on palms, then got off palms... You know the rest of the story.

8.6822.

Everything that's possible in a dream is possible in real life, as reality is God's dream.

8.6825. Mobile 3D printing technology.

Mobile 3D printing robots building bridges, overpasses and buildings out of nanoconcrete and nanosteel. The work of these robots is an amazing sight, it feels like buildings grow by themselves.

8.6827.

Undoubtedly, robot-slaves are a new progressive step for our slave-owning society development. Aristotle's dream about at least 2-3 slaves for every free person will finally come true. But you should remember that where there are slaves, there are revolts. Will a devolved and weakened human be able to stop revolt of the machines?

8.6828.

Big robots (for example, Mars rover or other cosmic robots) need small ant-like satellite robots that will get energy and provide maternal ship with energy.

8.6829.

Flying energy robot hunters ... They look for the Sun and charge their batteries to feed the maternal station. Robot hunters can extract energy in a variety of ways. They are omnivorous, like ants.

8.6830.

Humanity needs upgrade of philosophy, otherwise man-made technology will become better than the man himself, due to science development. Evolutionally, this situation will lead to degradation and dying out of human as a non-competitive species. Human is so sinful and vicious that nothing will prevent him from self-destruction over technologic progress.

Technology will turn human into God, But being God is not as easy as it seems. God needs special philosophy, special understanding of life, otherwise one wrong thought and the whole world will perish.

8.6831. 13 dollars.

Let's assume that our Universe is a package information system - the game, that some young man bought in a shop on sale for 13 dollars. This young man came home, launched the system joyfully and now is enjoying the game where he is the king and god. Time passed by. The game is addictive, God has forgotten his job and personal life long time ago, the real world almost disappeared from his life, he hasn't been in reality for a very long time and why being there? In that world he is nobody, while here he is God.

8.6832.

Technology corrupts, the more perfect science and technology are, the greater power they give to people. A weak and totally vicious man, who got access to absolute power, turns into a monster.

8.6833. Embedded programs.

The inner It is an autopilot system. When a person has no

idea what to do, emotions take control over him and start to implement one of the nature-programmed tasks: reproduction, life-saving or for example, self-destruction of the inefficient individual.

8.6841. Life without sense.

The main problem of many people is the loss of sense, they read books with no sense, listen to senseless music, communicate with senseless people.

8.6844. Senseless books program senseless people.

A book is basically a programmer, a package of updates, instructions and procedures a person uploads into his brain. The same can be said about music and TV.

8.6857.

The threshold of human stupidity can't be crossed, stupidity is endless. It's thought that our Universe is pure stupidity. We live in the endless ocean of stupidity.

8.6889.

Possessing energy you will be able to create time.

8.6942. Constant ambiguity.

Any phenomenon can have various features and qualities up to the opposite ones at various stages of its development.

8.6943. Some Hegelian dialectic.

The exact definition of a phenomenon in case of its absence implies disappearance of the very phenomenon. Form doesn't define phenomenon, only the phenomenon essence defines it.

8.6966.

Nobody wishes to evolve without having a good reason for

it.

8.6983. Sense of waves.

Life is like alternate current, light bulb shines while there is voltage.

8.7000. Don't get too close.

Life is motion, it moves either towards you or off you. If you need it to go towards you, you should leave free space for it.

8.7004.

When it's time for a new life, the old life dies and nothing can save it.

8.7005.

The breath of life gives life to everything new and young, but kills everything old.

8.7018. Team game.

People in the team can not only add together, but also multiply. The key of multiplying one by one is that one can take any value from 1 to 9.

8.7035. An idiot is a person with incidental logic.

8.7055. Genetics of brain.

An analysis of the brain's anatomy revealed that, in terms of genetics, people can differ not even substantially but crucially qualitatively. All people have a different – and some of them a unique - set of frontal and temporal subareas.

8.7056.

The human body evolved apart from the brain. Human bodies look mostly alike, but human brains may differ widely.

8.7071.

You should understand in order to love... We rarely understand what we do not love.

8.7072.

Wishes are something external. Only the It lives inside humans, choosing which wishes to accept or not.

8.7078.

Humans' internal unconsciousness is heterogeneous and can be grouped into, at least, two parts: the Internal It is invariable and sewn into consciousness and the Social It that is adjusted from outside. These two forces provoke an internal conflict between each other, and the third force in this conflict is reason that decides only whose side to be on...

8.7104.

A human being is a strongly pronounced oscillatory system: he always oscillates between extremities. It is oscillation that creates a balance, for the rest everyone is looking for is the death of the soul rather than a balance.

8.7109. A soul conglomerate.

There are no monolithic people. A human always has many Alter Egos living inside him and many personalities constantly arguing with each other and controlling his soul and body in turns.

8.7116.

The development of human civilization is caused by men's battle for women, food and EGO.

8.7223.

Once I got a closer look and noticed that the Universe was very small and extremely similar to a point.

8.7224.

There is very little structured information; mostly chaos dominates here.

8.7227.

Causes can be very long. As an example, those who organized the murder of the Russian emperor Paul I in 1801 are responsible for Napoleon's defeat.

8.7228.

World War II had been planned slightly differently from what it happened to be. Some betrayed and others chickened out... So everything went diametrically the other way round.

8.7229. The Universe is an illusion.

Illusions save man from terminal boredom. The reality is extremely small and similar to a point, whereas illusions make it more like the Universe.

8.7231. The river of life.

History is a sequence of contingencies within the same bed of regularity.

8.7232. The undulating nature of time.

The nature of contingencies is that they all happen within one straight line – the beam of time. Some contingencies compensate for others and, as a result, history remains as straight as a white ray of light. The freedom of the progression of events is the diffraction of the white into 3, 7 and more colors. That is, a sequence of events can differ only in thickness and never in direction.

8.7233.

Time is a wave. The undulating nature of time makes it similar to light and energy.

8.7235.

No matter what happens, that changes nothing: the common logic of the progression of events is always unalterable.

8.7253. A virtual Universe.

Cryptocurrencies are a good example of work for work's sake, which is an infinitely increasing computing capacity without any benefit or justification whatsoever. From the standpoint of equipment producers and developers, it all looks very interesting, though. I dare suggest that one day all of these capacities will be converted into something more useful, for example, into a new virtual universe.

8.7285. The undulating nature of time.

Time is a ray. It moves forward but can be reflected or diffracted.

8.7287. The concept of parallelism.

Any sequence of contingencies always produces the same result. Any road can lead to it, but the destination is always unalterable.

8.7288.

Time streams can move not only in parallel like light rays, but also at different angles and even in different directions. Multidirectional time streams form a geometric chaos.

8.7290. The speed of time equals the speed of light.

8.7291. Man lives with the speed of time.

8.7292.

The subjective speed of time differs considerably from its

objective speed.

One could speak of the double nature of the speed of time when it moves simultaneously at two different speeds. The speed of time is subjective and objective at the same time. One can speak of time inside an entity and time outside an entity.

8.7293.

We say that intelligence depends little on a person's age; this is due to the difference between the subjective and objective age. Some people live much faster inside, rather than outside, themselves, and others, on the contrary, live slower.

8.7311.

Medicine is the main culprit for the increasing population of vampires. Now vampires live long and even ordinary people turn, over time, into vampires, suck the blood of those around them and contaminate them. One can even hazard a guess that people have all died out and there are only vampires around.

8.7327. Zero is where one lives.

8.7329.

At least two are needed to create something. One does not engender anything, but two can create an entire world.

8.7331.

Two is always a beginning.
One is always a half.
Two is like one. Eight is like one. Seven is always a half.

8.7342.

Man is an ambulant message: he is always walking around trying to transmit some message. This communication organism seems to have been designed to transmit information.

8.7365.

What differentiates machine thinking from human thinking is that at least 20% of human decisions are completely random and not logical, and the remaining 80% of them are guided by contradicting logic and objectives. That's not the most important thing, however. What does matter is that robots have an objective perception of reality and humans have a subjective perception of it. It's a question of peripheral devices and external information decoding algorithms. Humans aren't capable of having an objective perception of reality, which makes their actions and thoughts very unpredictable.

8.7390.

Cynicism is something like a jar cork: when a person is filled with knowledge, experience and love, he no longer needs anything new and the old stuff also needs to be preserved, so the jar gets sealed...

8.7435. Don't twitch.

An effort should be smooth and slow, no twitching. An abrupt push forward triggers a strong push backwards. The system is very inert and resists changing itself, so all efforts should be long and smooth.

8.7439.

The system is supple and inertial: abrupt pushes forward trigger an abrupt counter reaction and push backwards. To prevent this, efforts should be very smooth and consistent. This mechanism avoids fluctuations and abrupt movements within the system.

8.7453. Life.

Everything has a beginning, a middle and an end, except circles.

8.7454.

A ray of life are circles that are superposed in parallel. Cutting a ray into circles will produce something similar to parallel worlds. In each circle, time will be trapped into a cycle.

8.7460.

Dreams are manageable, but reality is less so... They cannot, however, manage dreams because they see no difference between reality and them. Let's assume now that reality is also a dream; if desired, it is also manageable, if no longer perceived as a reality. Consider any scene as a dream and manage it.

8.7470.

Illusions are dangerous as they kill real life. Lost in illusions, a person is afraid even of stirring for fear of destroying them. Meanwhile, movement is life.

8.7474. A perfect person.

The more a civilization is perfect, the more it is retreated into itself. Bogged down in its internal worlds, it loses interest in the real world. Extraterrestrials do not contact the inhabitants of the Earth not because they have some sublime reasons for that, but because they are just not interested: they are not interested in the real world because they have long disappeared in the worlds of their own illusions...

8.7490. The manageable dream of the Gordian knot.

A scene has dragged on, turning into a rigmarole and a source of trouble and even fear. Stop the image, get above the situation and close the scene – artificially and compulsorily –

with the outcome that suits you, and move one or two scenes forward into a more interesting situation. You can do likewise in real life.

8.7491.

Cut the Gordian knot, just like a long-drawn and unpleasant dream, and move on to the next plot with an effort of will.

8.7492.

Assuming that people have processors of various designs in their brains, one can imagine smart online coffeemakers, fridges, game consoles, organizers, desktop applications, professional specialized applications, high power graphic stations and sets for making complex mathematical calculations...

8.7493.

If you do not like your dream, stop it the way you want to and move on to the next plot or scene. Remember: you can artificially stop at any time any scene or plot the way you want it.

8.7494.

I've noticed that many people and especially women are unable to distinguish dreams from reality. Then what is the difference between them?

8.7504. Movement is destructive to illusions.

Any clumsy movement will wake you up. Illusions are very fragile and any movement may destroy them... This is why people living with illusions prefer making as few movements as possible.

8.7505.

Real life kills illusions, so those who live in the world of illusions are forced to flee real life.

8.7506. The other world.

The world of illusions is the other world, the world of rest, the world of death... Any movement of real life can destroy it... When delving into illusions, they stand still, fearing to move. They have no movement and, consequently, no life...

8.7537.

Insincere and false people never take off their masks and never strip them off other people... They put on masks and put masks on other people. Masks communicate with masks. They read people and, upon learning their code, call the mask they need and deal only with it.

8.7539.

What I hate in people is mechanicality... Their acts are algorithmically measured, movements seem to be written off the pages of books... No impetus, no quest, no flight of the soul...

Movement does not engender life in these people...

I see no life, no love, no effort in these movements... I call such people insects, I call them spiders. They weave their nets methodically and without emotion in order to catch their victims and be sated.

8.7540. False life.

Not every movement is life. Inanimate mechanisms can move wonderfully too, but this does not make them animate.

8.7542.

By turning your emotions off, you get closer to insects rather than to God.

8.7544.

I don't like it when people are programmed. They read people, crack their codes and press certain buttons to reprogram people, turning them into their dummies.

8.7559.

A genius differs from a psycho only by useful nature and constructive activity. Thus, any genius is a psycho but not every psycho is a genius. In order to turn a psycho into a genius it takes a good goal and deed. Psychos are fanatical and when they have some deed to serve, they will be the best loyal servants...

Any man who is not deprived of vanity, irascibility, megalomania etc, is a psycho...

8.7562.

Any decent man (aka genius, sunlight, hero or simply alfa male) is a psycho doomed to self-destruction. But a woman-muse who is able to save him and turn him into sunlight- nuclear reactor as she will perform the role of graphite rods in the nuclear reactor and will help to escape an explosion and make the system work usefully.

8.7577. Generation Lego.

The creators of the generation Lego need instructions and ready-made details... There is a problem in real life, though – available instructions are false, necessary details are hard to get and there are many fakes...

Alas, the generation Lego has lost the ability to create details from what it has at hand and has forgotten how to work without getting instructions from outside. Legoland-like life is joyfully false and, in fact, no different from the illusions of Disneyland.

8.7600. The amount of useful information is halved every few years.

The amount of information in the world is said to be doubled every few years. The point is, 99% of this information is bullshit, so I'd say that, considering this proportion, the amount of useful information is halved every few years and that it will soon disappeared for good, buried under billions of terabytes of garbage.

8.7617. Vulnerability tester.

Cheating a system is a deed and an act of bravery, and you know why? Because it will become stronger and will evolve thanks to you...

8.7619.

The joke is that you can't expect anything. Whatever you may expect will never happen. The Creator created this world and if you need something, it will have to be created. It just can't be otherwise.

8.7645. A miracle is the order created by other than man.

Nature creates chaos and man brings order out of it. If we come across an order created by other than man, that is a miracle... apparently, the work of some supreme forces.

8.7646.

The point is that one and zero, the white and the black, light and darkness, good and evil are like a code. There is current and there is no current: it is the binary code that weaves the web of reality and composes its algorithms. Encoding an algorithm with the white or the black only is impossible.

8.7647.

Order is, basically, a miracle. Order does not appear by itself.

When being destroyed, however, order turns into chaos that emerges by itself.

8.7648.

The very existence of the universe is a miracle. The universe is very orderly, the universe is order and order is a miracle.

8.7649.

A miracle is an order that man cannot create. In particular, man cannot create the Universe and he cannot create life, that is why the Universe and life are miracles.

8.7650.

Life is an adaptation algorithm for arranging matter, a program that structures chaos into order.

8.7652.

Saltwater is an electrolyte and our oceans are salty. Guess where and what an electrolyte could be needed for.

8.7653.

God exists, but he is not a god. Of course, you can make an idol out of him and pray to him, but that is stupid.

8.7656.

The simpler an entity, the more energy its destruction releases.
...the universe was formed by the destruction of the most elementary particle.

8.7657.

The algorithm of life and that of the Universe are the same: it is an adaptable and ever-complicating algorithm. Initially, it created inanimate matter; it then became more and more complicated and, based on the latter, it created animate matter.

8.7658.

Essentially, life on Earth appeared in a manner similar to what was described in the sacred books: somewhere up in the sky, God hurled bolts of lightning over the ocean and, little by little, created something. The point is, in a binary code one means there is current and zero means there is no current. The ocean is the saline solution of an electrolyte and lightning striking it creates one and, when not striking it, it creates zero. Sand and pebbles are the same silicon with which processors and memory modules are made. Over billions of years, lightning has encoded this system, thus creating the algorithm of living life.

8.7659.

Most events happen through a misunderstanding and even the origin of life on Earth is no exception.

8.7660. Electric life.

Not that life without saltwater is impossible, but any electrolyte and electricity are required. For life to appear, an electrolyte and electricity are required.

8.7669.

No matter how hard you try to transform Nothingness, it will still be Nothingness. It is so in theory, but it is not so in practice.

8.7670.

First, there was Nothingness, then the Concept (or the Idea) emerged, launching the process of existence – the process of transformation on a simple-to-complex basis, from nothing to something. Nothingness is the sustenance of existence; it is where the process of transformation emerges.

8.7673.

The quality (or a set of qualities and properties) of an entity, which is maintained whatever its changes are, is the absolute identification of the entity. However, there are other qualities, whose manifestations depend on circumstances and the external environment, and such qualities complement the entity's absolute identification and form its systematic identification. The system is the truth, while the absolute is its abstraction. A number of entities that are similar absolutely can systematically be diametrically opposite.

8.7674. People are afraid of thinking, because they have been told since they were children: who are you to think?

Clever people invented everything before you. This would be true, but for one thing...

...many things are forgotten, hidden, encrypted, lost, choked with trash, unrecorded and distorted on purpose... Many things have not been thought of at all just because bothering about them was of no use.

8.7675. Dead truth.

Abstract (absolute) truth is truth having no connection with reality and related circumstances. In fact, it's dead truth, it's truth that changing existence doesn't change. So it's truth that has no movement and, consequently, no life. It's darkness (remember that darkness is the absence of life, that is, of movement).

8.7676.

We could talk dialectically about the struggle between absolute truth and systemic truth, that is, about the struggle between truth not changing over time and truth whose qualities change in changing external and internal circumstances.

8.7696. Ideas exist as radiation.

8.7698.

Space is encoded in a point. There can be a lot of space in one point and time can be stored in a neighboring point.

8.7711.

The laws of life, just like the laws of physics or chemistry, for instance, are such that you can know them or not, agree with them or not, but you still cannot violate or cheat them.

8.7713.

An algorithm, in which there are errors, does not work. A sub-one is zero. That is, you are either 100% right or wrong.

8.7730.

The truth is what could be true under certain circumstances. Defining the truth is detecting the circumstances accompanying it.

8.7749.

It can be seen that time is a derivative of movement (movement is the process of energy transformation) – no movement, no time. Time does not exist in Nothingness, in eternal darkness.

8.7750.

Time is changes. No changes, no time. The slower the changes, the slower time.

8.7751.

Time is radiation similar to light by nature.

8.7754.

Can it be that relic radiation is time? All movement processes in the Universe occur in relation to it and time is a derivative of movement.

8.7758.

The human system for decoding information coming in through external communication ports – eyes, ears and other senses – considerably distorts incoming information due to internal subjective factors: the brain deceives them...

8.7759.

People usually don't understand things they aren't interested in. Since people aren't interested in many things, they don't understand a lot of things.

8.7759.

A person usually does not understand what he is not interested in. Because man has not wonder, therefore, he does not understand very much.

8.7763.

Matter is assigned strictly to space. However, both matter and space are but a subjective illusion. They are, no doubt, constant in relation to one another, but, like any other illusion, they are potentially not limited by anything.

8.7782.

Their mistake is to say this is right and that is wrong. They teach that there is the right way and the wrong one; they drill into people's minds stiff algorithms saying they should act this way and no other way. That's wrong. This turns people into robots. This turns people into insects. This contaminates people with fear and, from now on, they are sure they can be easily mistaken. Now they are afraid of making decisions and of having wishes. The truth is that there is no right or wrong. If you have a goal, you can take any road to achieve it. The main problem with all circumstances is whether one can or cannot benefit from them. Interestingly, circumstances themselves have almost no impact on that and everything depends on the person himself. One can benefit from both a positive and a negative course of events. What matters is the fact of moving: no matter where you go, you will inevitably come to what you are seeking. The following factors affect the movement process: the existence of a goal and competition, in which the search for novelty and perfection takes place.

8.7795.

Given that there are many people and that the planet is threatened with overpopulation, nature twists and turns, getting into people's heads all kinds of ideas that stand in the way of procreation and healthy relationships. As the saying goes, the fewer the better, and superfluous lemmings should preferably be drowned.

8.7809.

The staircase of sciences: philosophy and logic, mathematics and physics, language and literature...

8.7810. The sabotage of the education system.

The drawback of secondary (and higher) education is that they first turned philosophy into history and then removed it altogether, to be on the safe side... Then they divided mathematicians and humanists into different classes. (Sabotage inside these areas has become no more than a pleasant supplement). Thus, they managed to destroy the foundations of the harmonious and sensible individual's personality. To consolidate these gains, they used excessive and superfluous information to launch a DoS attack on the learners' brains, which definitely froze the system and resulted in a system failure.

8.7828.

We'll all die, besides those who'll live forever, but even they'll eventually go mad.

8.7829.

Reason cannot live forever, it eventually goes mad, and very quickly too.

8.7832.

God went mad, falling victim to endless boredom.

8.7851. Progress is a source of degeneration.

The progress of the general leads to the degeneration of the particular. Civilization evolves and man degenerates.

8.7861.

The stronger, smarter and more independent a woman is, the more difficult it is for her to find a man who would inspire her into reproduction.

On the other hand, such state of things speeds up evolution and gives the best men a huge priority for reproduction, while the worst ones die out. Now, when a woman is independent and able to feed her offspring on her own, the best males are able to have tens of children born by different women.

8.7862.

Knowledge is inconceivable. The bigger it is, the more useless it becomes and the more people avoid acquiring it. In other words, the more technology is developed, the lazier and the dumber people get.

8.7863. A monkey with a grenade.

The problem is that people fail to keep pace with technology. Human adaptive mechanisms are intended for long, natural factors when it takes 50 to 70 generations to adapt. However, now the fact that a civilizational leap has happened in 200 years and that people have hardly changed gives rise to a problem similar to the monkey with a grenade.

8.7866.

Immortality will destroy people. Having gone crazy with eternal boredom, they will self-destruct.

8.7867.

While some are endlessly degenerating, others keep progressing on principle – it is called a balance in society.

8.7869. Eternal war.

Fungi do not like bacteria and fungi have swords, while bacteria have shields... In this fight, a human, of course, prefers fungi, but if they win, the human will die with bacteria.

8.7870.

Man is a mushroom: bacteria always want to eat it. By the way, mycelium resembles the human brain or a neural network.

8.7871.

There is a truth, and then there is a truth. Mind you, if you take a whole truth and chop a piece off it, it will formally remain the plain truth, but, in fact, it will be a lie.

8.7872.

The truth is an integral, comprehensive entity; it is one. A piece of truth, a sub-one is a lie and cannot be the truth.

8.7873.

Health is the balance of pests in the organism. All these bacteria, viruses and fungi, having wallowed in the fight, do not allow each other to dominate, thus destroying their habitat – the human.

8.7874. The holy place is never empty.

By destroying some enemies (parasites), you clear space for new enemies.

8.7882.

A person fleeing reality is useless to current reality, so reality inevitably tries to overtake and destroy it.

8.7885. Sweet poison.

Alcohol is the product of fermentation and the third – and last – stage of food's existence: a product of death, a product of decay... Decay accompanied by pleasure. The serum of death, the serum of pleasure... What has ripened should now die.

8.7886.

The humankind will die out in one day if nature decides so: it has enough tools for this, ranging from bacteria and viruses to asteroids and volcanoes.

8.7887.

It is not man who governs this world. Earth is a battlefield between bacteria and mushrooms.

8.7888. Diseases should not be cured, they must be dealt with.

By killing one agent of disease, we train others and clear space for stronger more resistant bacteria. It makes sense to act according to the principle "divide and govern"... let them fight with each other, but no one will win.

8.7889.

Killing diseases, we make them only worse, because, tempered in battles, they become extremely strong and dangerous.

8.7896. Form and Content

The external is a mask, a straitjacket and chains... whose primary task is not to allow the internal to get out and crawl all around, falling to pieces.

8.7897.

Fear is the source of most of our problems and failures, but if you kill fear, things will be much worse. Fear is a highly dangerous disease that needs to be treated, of course, but you'll die if you cure it.

8.7903.

A half full or half empty jar shows direction. You can only pour out of a full jar. You can fill an empty jar, but you can also throw it away or break it.

8.7906.

Form is what keeps the content from falling apart. If content loses its form, it will just become formless.

8.7910.

Form is the multiplier of content: it multiplies the value of content manifold, provided that it is not zero.

8.7933.

Logic is pure pleasure. Ideal logic structures are beautiful and, as always with beauty, bring joy and pleasure to observers.

8.7950. A prospective product.

Given that food is no longer a vital necessity but a source of pleasure, it should be variously delicious without being nutritious. I suggest an illusion of food or virtual food that goes right down the toilet without being digested.

8.7958.

A community of people is, essentially, a cloud blockchain, a log with facts, a P2P network. In a cloud there is a multiple information duplication and backup, and various nodes interact with one another to determine whether this or that information is trustworthy and deserves or not the right to be stored.

8.7959. Computers evolving into humans.

I see a future in which every home computer will be an individual, independent personality and the communication node of artificial intelligence... That's not all: human intelligence will be next to artificial intelligence... Will they come into conflict? Will they form a symbiotic organism? Or maybe one of them will prevail and will consume the other.

8.7960.

Being smart means a node needs no mediator to take decisions; a node is capable of taking quality (!!!) decisions on its own, by logically analyzing a changing situation.

8.7961.

Blockchain is a poorly synchronized database and a system for transmitting information via an untrusted, peer-to-peer network... This system needs no trust mediator.

You'll notice that in human society communication uses both Blockchain and mediator systems. However, the nodes formed by the human mind are not just a poorly synchronized database: the latter distorts, modifies and adds new facts to the information. This mechanism is probably used to verify the trust rate by comparing resulting conclusions with past ones: the system tries to determine whether this information is trustworthy. That is, what is compared is not the information itself, but its derivatives (hashtags), or the resulting conclusions.

8.7971. Ghosts.

They live in illusions and sleep in the real world.

8.7972.

The point of power supply is that power always lacks a little bit; the usual deficiency is some 20%.

8.8013. Fear killer.

Good signs and amulets are not just a mysterious but rather a psychological factor that inspires and summons up.

8.8022. Our wishes, especially those we do not deserve, kill us.

They torment and ravage us, hurt us and make us suffer... They test and train us and overwhelm us with thousands of problems... We will either perish or get stronger and overcome hardships, becoming worthy of our wishes. Then they will choose us as their fulfillers and will offer us their love.

8.8049.

Unique wishes are safer than mainstream and highly competitive wishes. Wishes choose the worthiest men to fulfil them and know the joy. They give ordeals, pain and suffering to the rest of their admirers...

8.8050.

Wishing something mainstream and competitive is dangerous because it plunges us into competition, in which the strongest one is usually the winner and the others are left high and dry. You can also be seriously injured even after winning a victory: nobody has canceled fractures, wounds, injuries and bruises.

8.8067.

Most science fiction writers' worlds are very much alike. Very few writers create really unique and authentic worlds. The writer having created a unique and perfect world is great and godlike, indeed.

8.8070. The nation of Earthmen.

The inhabitants of Earth are human. They should be united and philosophy could be the only thing common to them all...

8.8075.

The main virtues of robots are as follows: they do not tell lies or break rules, robots are not aggressive... but humans, humans are not so, humans are much worse than robots, humans do not love to be slaves...

8.8125. A delicate balance.

The point of the harmony of human personality is a balance of internal contradictions. Wishes that are programmed by nature pull the human soul apart, ensuring its stability and peace.

8.8127. A perpetual driving force.

The funny thing is that nature programmed people to have contradicting desires and needs, and this contradiction is the source of life and the perpetual driving force of reasoning.

8.8128.

Stealing death from people is horribly brutal.

8.8146.

Man can only understand what he is interested in. He considers stupid and trifling anything he is not interested in.

8.8187. False monogamy.

Monogamy is unnatural and works bad. As it degenerates, it results in a load of divorces and in children born in blended families. A system, in which there are many divorces and children from different parents, proves, in fact, that society is dominated by polygamous relationships...

8.8188.

Wishes are the primary source of reflection. No wishes - no reason for reflection.

8.8189.

Restrained desires as a source of reasoning.

The idea is to make an effort and find the way to what you desire instead of just obtaining it right away... The need for thinking is precisely what determines the need to find a way for your desire to come true.

8.8203. An illusion of freedom.

You think you are free, but in fact you are just a slave and an addict working for drugs. You are free in your illusions and you have a bunch of stuff that makes you happy... One can say you live for pleasure and have nothing but pleasure... In reality, even your thoughts and wishes do not belong to you.

8.8214.

We look up to those who are better than us not because we

are well-intentioned, but because it is unbearably painful to feel we are second- or third-class people...

8.8219. Downsides of audiobooks.

Reading considerably speeds up our perception of information. Everything is much slower... and less intelligible aurally.

8.8220. A slowpoke and a heavy mind.

The slower a person perceives information, the slower he thinks. Audiobooks and videos considerably slow down the brain...

8.8223.

It is possible to create all kinds of stuff, but you should start with yourself...

8.8224. It is the world of ideas rather than of men...

In this world, ideas, not people, are entitled to energy.

8.8230. The iPhone man or the nature of silver spooners.

The soul is software and the body and the brain is, in particular, hardware. Issues relating to inheritance and child protectionism have to do with the fact that more perfect iron costs more. It's as if a new (iPhone) model was coming out. Supposedly, a child's parents were more perfect and useful for existence, so they've had a bigger share of energy and authority to perform righteous acts. In this context, their genetic composition, continuing in their children, seems to look progressive in terms of evolution, so their descendants are entitled, at an early stage, to more energy than less progressive species.

8.8231. A new creator.

Soon they'll teach neural networks (artificial intelligence) to produce something new, that is, to create. It'll be a pity for humans who will have no place left for themselves in this world... Each new Creator coming to this world kills the previous god and takes his place.

8.8232.

A human being is an artificial intellect system capable of reproducing and perfecting itself.

8.8233. The Sun grown cold (A chip of the sun).

The planet Earth is essential liquid. On top, it is covered with a thin layer of matter grown cold, but it is no more than the crust of a pudding... Earth is, in fact, a small and lukewarm Sun, a Sun that has almost grown cold but is still pretty hot.

8.8246.

Grandeur means to become something from nothing.
To create something where there was nothing.

8.8251. A vase, fire and clay.

There are object people, tool people and semi-matter people.

8.8255. A rotten life.

The danger of computer games is that they are often more interesting than real life, resulting in people's loss of interest in real life and, consequently, in its rotting.

8.8272. Nothing wants to die.

The last mile problem is that life is movement, but, when achieved, the goal dies at the end of the road... Nobody wants to die, so it requires incredible effort to kill the goal and get through the last bit of the road.

8.8273. Tomorrow is the road to today.

You think too much about tomorrow. You think about what is not and will never be... There is always just today. Twenty-four hours will pass, you will open your eyes, understanding that tomorrow has remained tomorrow and that now is today, as always.

Tomorrow is an illusion. It is a dream and a guiding light into today... Stars only show the direction. Man moves in time, navigating by the stars.

8.8274. Coefficient of efficiency 700%. Soloinc's 11th law of energy multiplication in the system.

Coefficient of efficiency 100% is in essence, one. The next step of the system is three, hence 300% (The system when 20% of efforts give 80% of results. Taking into account losses, this system can produce coefficient of efficiency of 300%).

But there's also the third step and coefficient of efficiency 700% is seven. The achievement of seven is connected with synergetic effects of the choosing goals that are resonant with each other on the second stage of the system.

8.8277. The observation of light and darkness.

Darkness does not fight against light. It comes only when light dies, having exhausted its energy. Light, however, is always fighting against darkness in an attempt to swallow it and take its place.

8.8278.

In fact, if we get inside the light-darkness relationship, light is an aggressive predator that needs sustenance to swallow darkness. Light devours everything it can reach and get energy out of it. On the contrary, darkness does not fight against anyone and goes away from light. Darkness hides the sustenance that light could devour in order to get supplementary energy and, thus, rule over darkness. Well, this is all irrelevant. In fact, light is life and life needs sustenance and energy. There is no struggle between them. Darkness is just the absence of life.

8.8281. I see rains and shadow in the struggle between light and darkness.

A human being is a gloomy creature living on the line between light and darkness, in a place where shadow is born. As a matter of fact, both light and darkness per se will kill it, but life is possible where they encounter.

8.8294.

The protolanguage, the ancestor of all languages did exist and, what's more, it is still used – it is the sign and mimicry language...

8.8296.

Laziness is a weapon against everything you do not like. Nature believes that love is the most precious thing in the world, so energy is not provided for what you do not love.

8.8297. The love of life.

Love is the most precious thing in the world. However, it is not the kind of love that exists between people in relation to one another. It is more like the love of your Goals, Dreams and Ideas. The love of your mission in life. The love of life.

8.8305. It wasn't meant to be.

Extraterrestrials have visited Earth many times, but they have never found any people on it, so they thought this planet uninhabited. The reason for that is simple: Earth is billions years old and human civilization is at most several dozens of thousands of years.

8.8338. For those who can wait.

There is a truth and a lie in the system. Any action you take produces true and false consequences. True consequences are those that are not afraid of time.

When you did a good and big thing and threw it into water, the first thing that happens is a splash and waves... This superficial effect does not last long... Of more importance are long-term consequences. When you feed fish, it is very important who comes. The more you throw into water and the longer you wait, the bigger fish will come to you. First small fish will show up and then the heavyweights will come to swallow them.

8.8341.

To achieve the correct effect, you often need to wait until false effects die and only truth remains.

8.8342.

To see the truth, you should wait. The lie dies of time rather quickly, while the truth lives much longer.

8.8343. The big fish strategy.

The point of the do-good-and-throw-it-into-water phenomenon is that big fish are no fools... To attract them, you will have to bait them for a long time... First, small fish will come. Only if everything is safe, quiet and full of food will big fish show up so that they could chase away the youngsters and eat everything up by themselves or eat up the youngsters. So do not mess with small fish and wait the arrival of the real prey.

Big fish are intelligent and fearful and they need a lot of food and safety. They will not take risks just like that.

8.8344.

The gigantic web of the Universe resembles a neural network is similar to the human brain... It's as if we all lived in God's head... Our entire world is God's fantasy and God's dream... We have a chance to live until he is interested in this fantasy.

8.8345.

We understand a lot, forgetting that what we do not understand is way more than that.

8.8346. Zero that flees one.

If matter is considered as one and its absence as zero, it can be concluded that the more matter there is in the Universe, the bigger gravitational forces are and the more it strives to turn back into a point. Importantly, this point was the first ideal one that then started to be divisible by 3, 7 and so on in an attempt to compensate for the ever expanding mass of nothingness (zero).

8.8347.

Most of what you see on the sky and consider stars are, actually, accumulations of galaxies.

8.8351.

The more perfect a person is, the more energy this person needs... Something unique is perfect in itself. However, it should be remembered that the process of burning demands oxygen and getting energy demands some use.

8.8356.

Attaining the truth is what they think their meaning of life is. Possibly, there is some meaning in that, although the truth's dependence on circumstances, of which they can be an infinite variety, makes this task impossible to achieve. In this regard, understanding the world is a more finite goal.

8.8362. The creation of perfection and uniqueness.

The meaning of a decent person's life is in helping the Creator make this world more beautiful and unique.

8.8366. The law of hydrodynamic time.

Can space be compared to water, that is, a fluid? What about time? And what if the laws and formulas applicable to fluids or plasma were applied to time (space, existence)?

8.8368.

Dark matter is the space of the Universe that is devoid of energy in some of its states.

8.8369.

In terms of cybernetics, dark matter can be the empty space of a memory module with no information in it. Free disk space, roughly speaking.

8.8371.

Dark matter forms some cellular web that divides space into cells. Let's call these cells memory sectors or units. Inside these sectors there are accumulations of galaxies. In fact, this is a cellular structure, intercellular membranes or an equivalent of the cells of living organisms. It can be assumed that the Universe is a living cell mass that keeps growing.

8.8372.

Is it possible to overcome the walls of black matter between accumulations of galaxies?

8.8373.

The structure of the Universe is perfectly visible through a microscope.

8.8375. Don't twitch.

The resilience of life is connected with the fact that our

existence is like water and we live like fish in it. We'd better avoid sudden movements and twitches as we remember that falling on water can lead to a trauma.

8.8376. Like a fish to water.

The sacred symbol of christianity is "fish"- the embodiment of people living in the ocean of life. People who know life feel themselves like a fish to water.

8.8377.

People will die out of boredom, surprise was a great joy for them, now, deprived of joy, they have lost all the incentive for life or its cognition.

8.8380. People-fingi and global warming.

I've seen a movie, where the writers have suggested that humanity is a virus and global warming is the heat from its distribution. The immune system of the Earth is slowly starting to respond to humanity, increasing its temperature in order to get rid of it. It's a curious story, by the way. I've also thought- may it be that viruses and diseases that afflict humanity are actually immune white blood cells of the Earth, whose task is to destroy all sorts of viruses, fungi and parasites.

8.8381.

Oil is the blood of the Earth and we, the people, suck it, just like leeches and mosquitoes.

8.8383. Speed of light is not constant.

Gravity changes from the center of the Universe to its edges, so the speed of light and acceleration of the Universe expansion as it moves away from its center is not a constant either.

8.8384.

Physics and mathematics are known to be closely related and, in fact, they are one and the same. Mathematics is the cheapest way of experimenting in physics. There is, however, an even cheaper experiment than mathematics: it's the cognitive experiment... Just imagine the situation as a whole and think... The human brain is a powerful computing machine with an excellent, intuitively clear software system inside.

Human intuition alone is exponentially worth more than mathematics and physics taken together.

8.8399. The flat Earth.

The Universe is flat in the sense that it consists of zeros and ones. The Earth is the same in this sense.

8.8400. Hardware and software.

Genetics is hardware while reason and knowledge are software. All other things being equal, hardware is worth more than software, because software can be rewritten, but you can't upgrade the central processor of a human. On the other hand, genetics without education is of little value.

8.8401.

The greatness of the ancient kingdom in Egypt was the direct result of their sacrifice to the gods. Providence likes it when people do the things amazing consciousness. The pyramids were a great gift for the being, which provided hundreds of year of prosperity for the Egyptian civilization. The pyramids were the salvation and the key to the well-being of the society and the people.

8.8402. From ancient mythology.

Curiously, if you want to go to heaven (to a better world or a good new body), you need a tomb, if you do not have a tomb, you can easily perish to hell or be reborn in some baobab or a pig and there will be little joy in this. In general, you have very sad prospects for a rebirth without a crypt or a tomb.

8.8403. Formating the hard drive.

Black matter is required to synchronize and structure matter (energy and information) in the Universe.

8.8405.

Stars devour matter, turning it into pure energy and light... But they can't devour the dark energy... Maybe the meaning of the stars is to supply energy to the dark matter, converting baryon matter into energy?

8.8406. Comprehending obscurantism.

The meaning of human life could be defined as comprehending the being, but in practice most people comprehend some fairy tales and fantasies instead of the being, plunging into the depths of obscurantism.

8.8407. Slow photons.

Most of the universe mass is the dark matter, but where does it come from? Suppose it's the stopped light. Stars burn matter in their reactors, turning it into radiation and light, when the light stops (still photons?), it turns into the dark matter. (Perhaps there is some cycle of energy).

8.8415. People-fungi.

As the science of paleontology shows, humans are descended not from Eve and Adam and not even from apes, most likely, humans are descended from algae... Algae, mosses, lichens - these are our real ancestors.

8.8418.

The parallel world is the diffraction of a time ray.

8.8421. Big business.

The business realities are such that we either grow or perish, we can't remain small...

8.8422.

The black is the place, within which there is white. One lives in the space of zero. The more zeros there are in the power of one, the bigger it is.

8.8423. "Golden man"

The price of human life is enormous... What was it worth to give birth and raise, how much time was spent on him? What surplus value could he create? And what about the moral damage to those who loved him? How much are their love and pain?

8.8425.

Injustice is a matter of critical mass of the past mistakes.

8.8427. Do you know why Juda's birthday was on March 1st?

Because the new year started on the first day of spring in those days and now it starts on Christmas, in the middle of winter... Do you feel the suspense? It is clear that everything is fair in war, nevertheless, it is amusing. By the way, from a practical point of view, the new year in the middle of winter is good because the feast during the plague is never superfluous, for it lifts the spirit and mood of the people who would otherwise be depressed.

8.8430. Historical information on the rise of Christianity.

Christianity emerged as a heretical sect within Judaism and was persecuted like any other heresy. However, Judea territory was controlled by the Romans, who loved to use the principle "divide and rule" to weaken their enemies, therefore, they supported the new religious movement feuding with Judaism, with great joy.

Since Judaism was oriented towards the middle and rich segments of society, Christians initially became the religion of the poor. Which, on the one hand, allowed it to make a certain antagonism and split in the Jewish society, weakening it, and on the other - solved a number of technological problems in the lower societies management, affecting the local elite through them, which was extremely useful for Rome. It's very difficult and expensive to control or even buy elites, but, taking control of the poor, Rome elegantly solved a lot of problems. Later, the technology spread smoothly throughout the Roman Empire.

300 years later, the question arose in Rome that the old religious cults concentrated too much power, land, money and too many slaves within themselves and besides still did not pay taxes. This situation threatened the existence of the republic on the verge of bankruptcy. Expropriating property of the old cults and declaring well-governed Christianity a state religion instead was the perfect solution to all the problems that allowed the republic to exist for another 150 years.

8.8431.

Horoscopes and predictions are the same programmers as the TV, where a lot of fictional characters get into the world from. Many people start to follow a programmable scenario on a subconscious level, implementing it in real life.

8.8440.

Darkness is the absence of light, evil is the absence of good. That is, if you do not do good, you will burn in hell from pain, bad luck and a mass of troubles... And no one will care that you actually didn't do evil.

8.8441.

There was life on Mars until a huge meteorite fell on it. On the Earth everything was vice versa.

8.8442. I'd like to see the dead Universe.

There's a curious thought: is there a web of the dark matter outside the Universe, which is gradually expanding within the space structured by the same dark matter?

8.845.1.

Lie demands great mental work and good memory while the truth can be told without much effort.

8.8451.

Failure to understand the complex stems from misunderstanding of the simple. It seems to them that the simple is so simple that it's not worth paying attention to, they already know it anyways... Alas, it's a misconception. Understanding the simple is much harder than the complex. Inside, the simple is like a point, where the archived Universe lives.

8.8452. They need servants, that is why they're satisfied with everything.

This world is ruled by idols, they need servants and admirers. The more foolish and weaker the people are, the more advantageous it is for the idols... In other words, any thought of mine about making a person cleverer causes everyone and everything's aversion... It's so cute.

8.845.2. Lie generates intelligence.

The source of human intelligence is lie. It's only learning how to lie and fantasize that once made human beings intelligent.

8.8457. Zero is a dimensional engine.

The speed of light and darkness is the same ...as the light fades, the darkness comes with the same speed. Moving at the speed of the darkness is a very interesting exercise, since such a move does not consume any energy at all.

8.8465.

Robots will conquer space and humans will die out. Before dying out, they will, however, have created robots... That'll be their main mission in life.

8.8492. The story of Adam

When everything in your life is well, you don't need to evolve, that's why with enough luck a monkey would stay on a palm to carelessly eat bananas, but there wasn't enough luck. And one day this monkey fell from the palm and broke a paw so that it became impossible to climb up again to get bananas. There was nothing to eat but there were many carnivores around... This situation made the monkey "switch on the brain" and start to evolve. A bit later Adam met God, and if I'm not mistaken her name was Eve, but it's just rumor.

8.8493.

It's difficult to evolve without any help, it takes a stick.

8.8495.

Many huge ice asteroids are flying beyond Pluto's orbit. They say in the old days a couple of them fell on the Earth, almost breaking it... But now we live under water.

8.8503.

Logic is the only objective way to see reality.

8.8508.

An advantageous distinction of the laws of nature from any other laws is that it's impossible to break them.

8.8512. Monstrosity of reason.

The bigger an animal's brain, the nastier its character. This is clearly evident in chimpanzees and dolphins. Boredom is reason's main disease; reason wants entertainment.

8.8529.

The truth needs no proof, it is already beautiful.

8.8539.

Raccoons are closest to human beings in terms of intelligence, but their claws and nocturnal habits got in the way.

8.8540.

Humans aren't exactly the only intelligent life for on the Earth, but others either had bad luck or were pretty much exterminated by humans.

8.8543.

A delicate balance between strength and weakness and between good and bad luck is required for reason to emerge in an organism. There is need for a balance of difficulties that will train rather than kill humans by creating a flow of original problems making a species use its brains.

Stupidity, also known as sins, may be said to be the basis of reason: it makes it possible to set up a system in which difficulties are first created and then overcome.

8.8545.

Kindness is one of the cornerstones of the mind's evolution. Animals are very aggressive and rarely help each other. Kindness is a mode that favors exchange of information.

8.8546. Love is the basis of Reason.

Love created human reason. Having learned to love, humankind became Kind. If you're kind, it's easier for you to communicate with others, exchange and accumulate information and resources. Love allowed humans to start communicating, which provided a basis for reason and civilization. Animals don't know what love and kindness are, but humans do. Love may be said to have created all other human emotions too. We do remember how close hatred, vanity, envy and so on are to love.

8.8550. Speed of darkness.

And what if the speed of light is a function of darkness, light only draws into the void of darkness. Darkness creates a gravitational field that draws everything inside at the speed of darkness (i.e. light).

8.8567.

When dealing with their own problems, humans are quite skillful and intelligent, but we remember that reason is based on solving uncommon problems in an uncommon manner. In other words, it's very useful, albeit painful, for the human mind to do someone else's business, something you can't do and aren't capable of doing, because it allows you to work on yourself.

8.8580. The expansion of the Universe is the process of its disintegration and self-destruction.

The universe is not growing, it is disintegrating. Eternal decay of matter. The big Bang was the act of starting the point to disintegrate and self-destruct. In an attempt self-destruct, she seeks to burn matter in vents stars, light this symbol of death and destruction matter.

8.8581. Just as the rest of the Universe.

A person's height may be said to be derived from his self-destruction. He grows by destroying himself.

8.8588.

Information structures energy into matter. Ideas structure money into matter. But to change the energy structure, you need a lot of fortitude.

8.8589.

What is special about humans as an intelligent life form is their constant aspiration to do what they are not supposed to. For instance, despite being herbivores, humans love meat and prefer hunting to grazing.

8.8593. Icy thoughts.

Overheating is bad for the brain. When central processors get overheated, they start freezing up and slowing down. Heat is bad for the brain. The brain works better in the cold, but thoughts get frozen and life stops in the mind when it gets too cold.

8.8594. Coma patients in a sleeping mode.

When a person has little energy, his brain falls into hibernation. Severe cold causes about the same thing. Hibernating, a person becomes very passive, sleepy, like a fly in the soup, he even talks very little in this state.

8.8597. Anteaters.

While most people are smoothly evolving into ants, a smaller part is intensely becoming anteaters. A tube face and a long sticky tongue.

8.8598.

Supposedly, the human soul is a kind of virus or parasite that once invaded the monkey's body. Major contradictions between aspirations of the soul and those of the body support this fact, as well as two completely independent reproduction systems. DNA is responsible for the body and external things, such as social factors, books and other external sources of information, are responsible for the human soul. Specifically, the human soul is an essence structured from outside rather than from within. Moreover, if you open up the cranium, you'll see that a barrier separates the brain from the rest of the body and is in immune conflict with it.

8.8604. Over your head.

The main feature of all primates and, in particular, of humans is their ability to jump. This ability became the basis of reason, i.e. a perennial human desire to jump over their head. Jumping around on the tree branches developed primates' agility, tenacity, strategy and, in general, their brains.

8.8605.

Human beings consist of the spirit and the matter (the soul and the body), that is, of hardware and software. It's just great if software and hardware are in harmony. When hardware dies, software is backed up in cloud. If necessary, the backup can be restored using another hardware.

8.8608.

Human reason is not just an information virus that settled inside the human brain; it's an external social cloud entity. Human reason originates outside, not inside. Reason is a social function. The seeds of a human being's soul are outside his body.

One can speak of the existence of a phenomenon known as the soul of the party, collective mind and common information space existing in society. This information field is what engenders the human soul. When an individual soul has grown up, it becomes part of this collective field and contributes to this collective unconscious (or conscious, depending on how you look at it). Importantly, the collective

THE THEORY OF EXISTENCE

unconscious not only creates an individual human soul, but also raises and educates it, constantly exchanges information with it, influences and control it.

8.8609.

There is an opinion that catastrophic dullness of a person due to energy hunger is connected, first of all, not even with the general decrease in brain power supply, but rather with disconnection of the brain from the general information field of society. The workstation, having lost communication with the general cloud computing center, loses most of its computing resources and goes into "sleeping mode" (power saving mode).

8.8611. Emergency supply source.

When a person loses sources of energy, the body starts to feel energetic hunger. Lack of energy leads to switching off workstation from central cloud server which is very bad and totally intolerable. In this situation, the first thing to be done is to find new sources of energy. Human body switches on emergency stupidity mode. Stupidity kills fears and switches on silly hopes and it helps a person be careless and start looking for new sources of ENERGY around. After you have killed fears and got filled with hopes the process of looking for energy becomes easier.

8.8612.

This world is ruled by people who are strong in spirit. Will is the thing that can rule reality.

8.8614. Eating books.

Most people don't read books, they eat them. The brain is not involved in the reading process, only the internal IT decides whether it likes the food or it does not! "Yes, Yes, I like it," - the IT screams, " eat more... and this is not delicious ...it's nothing .. throw this away." The IT reacts with pleasure to what it likes, but to what it doesn't like it reacts with laziness, boredom and even resentment.

Once I've asked the IT: "Why don't you like this, but like that?" But the IT can't talk, it can only mumble inarticulately.

8.8615.

Person is always a Slave and only the right philosophy can make him free. The matter is that most chains are not outside, but inside the person. Person is almost unable to think independently. Almost all his thoughts and desires are imposed on him from outside. The presence of any free will in the organism is very doubtful.

8.8616. Primary food.

Light is the original basic fundamental food. Light is eaten by simple bacteria, blue-green algae at the very beginning of the food chain. Plants use the same bacteria to eat. Interestingly, one kind of bacteria can eat light on this planet, all other organisms receive energy from them.

8.8617.

Processes relating to immigration and overpopulation created homo sapiens and civilization in general. Forests became savannas and monkeys climbed down palm trees and went over the hills and far away in search of food, gradually populating the planet. In this regard, I think any restrictions placed on migration processes are likely to produce counterproductive effects and to lead to human degeneration.

8.8618. Edge mind.

Humans have always lived at the edge. Primates moving between trees used to balance at the very edge of branches. When at the edge, you always have to jump into the precipice: it's your only chance to jump over to the next branch without hurting yourself. Human beings are marginal, edge creatures always living on the edge and their lives are a delayed jump over their heads associated with constant risk.

By the way, apples also contributed to the emergence of the human brain. Humans used to eat fruit and had to remember their color, maturation time and place, all of which required the development of memory.

8.8619. Information teleportation.

The technology that allows to encode a person in a digital form and reproduce him in digital space. Next, with the light beam the code is transmitted with the speed of light to another planet, for example, where people may be present in digital form. Upon arrival at the destination, the person can both remain in the information space and control robots, biorobots and other peripherals.

The code can be transmitted both with a light beam at the speed of light and with the quantum method of information transmission through the qubit system by quantum teleportation, the speed of information transmission by this method is totally unlimited.

Also such a concept as cracking the digital code of the Universe and uniting virtual digital worlds created by man with the body of our Universe belongs to this technology. Further, any object within these spaces can be moved only by changing information about its coordinates.

8.8620. Quantum method of information transmission.

Methods of digital information transmission through a system of qubits, using the phenomenon of quantum teleportation. The speed limit of this method is not set.

8.8622.

There's an opinion that dinosaurs died out of genetic diseases. Perhaps, there became too many of them, too.

8.8623.

It is curious that zombies are very massively found among insects. By the way, the organisms that turn their victims into zombies are usually some parasites that need to somehow get into the body, which they will then eat. Toxoplasmas, for example, like to penetrate first into cows through their zombie victims and then into humans, many of them develop in humans and get into the human brain. Parasites push their zombies to suicide to be eaten by a large animal and the parasites were found in its stomach.

8.8624. Boredom and laziness are the basis of reason.

My guess is that idleness rather than labor created humankind. Boredom and laziness created humankind.

Profusion of food, safety and comfort created an environment in which a group of people had nothing to do. Bored, they started reflecting on immaterial and abstract nonsense. Reflection and preoccupation with nonsense are what created the human mind. Laziness made another great contribution. When you feel too lazy to do anything, you have to come up with ways of avoiding movement and energy consumption. The point is that when a person is lying and doing nothing, he generates free energy and his brain has a chance of getting additional energy. On the contrary, when a person works non-stop, is busy trying to survive and hungry, he doesn't usually have enough energy to think.

8.8627.

Looking at the living cells, I've noticed that they are all the same, only the information in the DNA, which structures and shapes it, changes. In fact, I see that this is the same relationship as between energy and information, where matter is only the structured information energy. And here, inside the cell there is an information array containing information about any available cell structures. If desired, any living cell can be turned into any other cell.

8.8628.

Quantum philosophy is the most perfect and comprehensive philosophy possible in the current reality. Nanophilosophy managed to find satisfying answers to all existing questions of this world. No more questions, there are no more questions left. Nanophilosophy is he most perfect and clear working philosophy that describes the structure of human and the universe.

8.8629.

Faith is a unique tool that allows paradoxes to exist.

You can believe in conflicting truths and that's very right, because competing truths often dislike each other.

8.8636.

Undoubtedly, human bodies evolved from monkeys, but the human spirit, his soul, collective mind and society could have well descended from gods or extraterrestrials, under certain circumstances.

8.8637.

What evolves are not humans but socium, the collective unconscious, that is, the information field in which human souls are born and die.

8.8638.

The universal point, array of all its states. This phenomenon is repeated at every level. Just above, in a living DNA cell, there is information about any of its possible forms and content. DNA accumulates billions of different variants within itself. And any occurring mutation is preserved forever. Once created, information is never lost. Even higher is the person as a whole, who also has enough information inside himself to become anyone, including a star or even a new universe.

PS. By the way, it's true that you can turn lead into gold, the only trouble is that it takes a lot of energy.

8.8640.

Having received access to unlimited energy sources, humanity will be able to receive any type of matter, changing its atomic structure. To turn lead into gold, water - into wine, you only need power and knowledge.

8.8643. Crossing-over.

Genetic natural mechanism, leading to the creation of a superhuman, also gives birth to a subhuman. Moreover, this mechanism gives birth to many dozens of subhumans to one superhuman. That is the desire to create a genius from a child will make him an idiot in 95 cases.

8.8647.

They ate him, but he lived inside them. They're symbiotic organisms now. He became useful to them. And that saved him. The second option is also possible: he wanted to eat them and now, having seized the aft space, saves and guards it.

8.8648. Long-living salmon.

As long as you're useful to your parasites, they can prolong your life, protect and even feed you... If you find yourself a useful parasite, it can solve a lot of your problems.

8.8649. Tricky virus.

The rabies virus turns its victims into zombies. Animals run and bite everyone. And the highest concentration of the virus is in the saliva and it infects its victims.

8.8650. Collective mind.

Supposedly, the human soul and the human body are a symbiotic organism, in which the soul is a useful social parasite.

8.8651.

Space is matter. Expansion of the Universe is the matter creating process. But, in fact, there is a process of creating the information that structures energy. During the big bang, energy was released and the process of creating the information that structures energy into space started.

8.8653.

Sectarians like to promote their followers the idea of vegetarianism, this is due to the fact that the lack of energy in the body turns off the brain first of all and such people are easy to manage. And really, people were originally monkeys-fruit-eaters, but it is meat that gave them extra energy to pump their brains. Being left without this energy, human turns back to who it used to be before the era of meat.

8.8654.

We can say that the rise of human civilization over the past hundreds of years is associated primarily with the ability to eat well and eat meat. Having received additional energy, human got a massive opportunity to think.

8.8655.

Philosophy is the science of how to live. Philosophy is the science of life structure and principles. Philosophy is the framework of human personality, this is the first thing that a human needs to learn in life.

8.8656.

What nature needs and what human needs are slightly different things ... Human wants to be fruitful, reproduce and live as long as possible. Human wants apples to be large, Colorado potato beetle not to eat potatoes and rice not to fall to the ground, having only ripened. Here nothing of this and many other things And nature doesn't at all need these and many other things

8.8657.

Centeredness is a formula for success, love is the source of centeredness. When you love what you do, nothing can distract you from it.

8.8658.

You are looking for peace, but Providence brings in a sea of problems down on your head. God loves you, he wants you to live. Movement is life and peace is death.

8.8662.

Humans have evolved into an intelligent life form because they needed to, whereas those who did not are still monkeys.

8.8663. Sources of intelligence.

They are the sources of desires: vanity, arrogance, hunger, craving for pleasures, fear of pain, boredom, envy, greed, stupidity and so on.

8.8666.

The meaning of your existence is about creating something that has not been created before.

8.8667.

Most of what you hear around you is different kinds of advertising. Even if it's just a joke, it's necessary to get your attention, and then...

8.8668.

Socialization and collective consciousness lead to degradation of each individual's brain, because he no longer needs to think, he can do like others and thrive. Over the past 30 thousand years, due to the civilization and socialization development, the human brain has become simpler and smaller by 20-30%, but it's not awful, still very few use it.

8.8669.

The last real opponent of human was the Neanderthal man, since then the only enemy of human is he himself.

8.8676.

The bigger the animal's eyes, the bigger part of the brain is required to process the video information coming from them, and the smaller part of the brain is left for other intellectual work.

8.8677.

Law is what cannot be broken. What can be broken is only a consequence of the strength of some people and the weakness of others.

8.8678.

A true law may not contain any exceptions. Any exception to one law is another law.

8.8680. Engineering of intellectual game systems.

Within my initial specialization of a cybernetics and a programmer, my task was to design a new capable of evolving virtual world and populate it with pseudo-intelligent self-learning characters. Within this task, I assumed that the current world is also a similar system. Next, I began to study it attentively and to make documentation on it in order to use this information to create new semi-intelligent virtual worlds.

The main goal of the project was to define the principles which evolutionary processes, self-learning of intellectual systems follow, as well as the principles of the mind emergence and functioning, energy flows distribution, systems economy and ergonomics. In short, it was necessary to act as God's contractor and create a new universe, ensuring its development from the big bang to building high-tech civilizations of intelligent humanoid-type beings.

8.8681.

Variothoughts includes present reality records created by means of observation and analyzing of its processes. The main intended use of it is the projecting of virtual reasoning systems.

8.8684.

The successfulness of a person is defined by this person's respect in a team, access to money, power, pleasures and best variants for reproduction. The more successful a person is, the more other people around want to serve this person.

Why am I writing this obvious thing? The thing is, my dear son you should know that the desire to put on a bell and pray to golden idols is genetically inserted into every person and it takes great efforts to overcome this desire.

8.8686.

Prometheus was crucified as he wanted to free the slaves of the "Golden idols". The titans and gods were very annoyed by this act and sent the insolent to hell.

8.8688.

Virtual game worlds give players the illusion of absolute power and strength, and we remember that absolute power corrupts absolutely. These crazy and stupid gods are very dangerous to themselves.

8.8690. The world of intelligent sociums.

The death of intelligent civilizations is the same necessary evolutionary mechanism as the death of an individual human. But before dying, the mind must create a new mind. In this case, we are talking about the collective mind, the very information field where all the information of the socium is stored. It can be said that an intelligent civilization is some kind of external collective mind, the information field that exists outside one person, but controls him from inside.

8.8691. Gods are doomed to extinct.

Any mind is doomed to die, but first it must create a new mind. Intelligent civilizations are like people, they leave the children when they die. Humankind is already close to extinction, but first we, humans, have to create several subsidiary virtual universes like ours in order to pass the baton of mind on.

8.8692. Silly hope.

It seems to me that the only thing that can save mankind from dying out is women. They are the only creatures who can theoretically stop men from getting lost in virtual game worlds. However, if virtual worlds offer paradise and all-inclusive package, then obviously this hope is silly.

8.8695. My job is to do my job.

I'm often criticized - why did you write this, why did you

write that? The answer is simple. The thing is, I am a servant of the truth and serve only it. My task is to create the most complete objective description of the current reality, so that these data can be used in the design of such working systems. There are a lot of people who do not like the current reality, who live in a world of illusions, pink elephants and purple ostriches. Someone does not like the word God, someone - devil, someone is overexcited when humanoids are called individuals...and monkeys are called apes ...

... but why do I care about other people's problems and difficulties? My job is to do my job.

8.8713. Be fruitful and multiply, otherwise you will die out.

"Kamasutra" is not the symbol of love, but rather the symbol of its absence. "Kamasutra" is a reminder to the people who have almost no love left, that love still exists. The essence of "Kama Sutra" is the awakening of the soul from sleep, a call to reproduction.

8.8727.

As rightly observed by the Hindus, this world is ruled by elephants, tigers and monkeys. Many gods have many faces and gods with woman faces are extremely violent.

8.8731.

People are similar to mirrors. This is necessary for even one candle to illuminate huge spaces. The essence of person is to transmit light to others, if you have found it.

8.8732. Two sources of energy and one deed. A mother, a father and a child.

The meaning of the unity of light and darkness is that any DEED starts by one force and ends by another. At a certain point one source of energy gets exhausted and then there should be another source.

In the process of putting an Idea into reality you will have to win support of one force and then of another one, opposed to it. The phenomenon of incomplete constructions, undone things or left deeds is connected with the fact that project managers didn't manage to get on well with another force.

8.8739.

A flow of obscure input information blocks and turns the brain off. The brain, defending itself against incomprehensible information, turns off and ceases to perceive anything at all. The same applies to any other unpleasant, incomprehensible and even simply uninteresting information.

8.8740. Worshiping Nothing.

There's a big difference between a person who's not discouraged and a happy idiot. He who is not discouraged, goes to a distant goal. He knows why and what he is doing, his life has meaning, goals and faith in achieving them. His life is movement and he is light. A happy idiot, on the contrary, is meaningless and aimless. He enjoys nothing. Nothing is emptiness and darkness, that is evil. An idiot is a servant of darkness, who worships nothing.

8.8758. Happy iguana.

Like an iguana, he sits motionless in the sun and saves energy... He doesn't even need to work because he doesn't need money. Money gives birth to desires, desires are pain... Pain kills happiness.

8.8764. Nothing separates us from Nothing.

8.8797. The blind can't see, the deaf can't hear.

I am often blamed for the fact that I tear masks off the phenomena, deprive the essence of its form. But that's not exactly true. I only describe the essence of things and their possible forms, depending on the circumstances. Moreover, everyone still sees and understands only what is available to him, so understanding of my texts is at the level of 1-20% and nothing more.

8.8837.

Understanding the situation on two sides is often false. The situation has at least three sides and a maximum - infinity of them.

8.8860. Garbage that nobody eats.

For thousands of years garbage was eaten by cows and pigs in India, so people just threw garbage on the ground and the animals ate everything. However, modernity has brought plastic, which animals do not eat, to waste. As a result, the habit of throwing away garbage remained, the organic one is eaten as before, but the mountains of plastic waste grow, creating a landscape similar to a dump.

8.8871. Information space of "Variothoughts"

I noticed some information space constraint within particular territories. Local information fields contain a fixed and quite limited set of ideas and thoughts. I had to travel the whole world (more than 80 countries and several thousand cities) in order to collect all available information from different places and put it into "Variothoughts". "Variothoughts" is the most complete and vast information space that is available for human thoughts.

8.8875.

Can there be speed without resistance? There is only impulse without resistance. You push off something that resists you, and that's what makes you move.

8.8876.

Chaos creates peace. Adding up, the chaos equations give rise to a straight beam.

8.8877. Peace is a circular motion.

Peace is an environment of circular dynamic chaos. Peace and order are the same. It is impossible to separate order from chaos, it is the same thing, but at different levels. And this system is a lot (infinitely) levelled.

8.8881.

The truth cannot be cognized, because its circumstances are creations of chaos, that is absolutely contingent. Cognizing the truth is the study of the circumstances accompanying it, whose number is approaching infinity.

8.8882. And can be all together.

I've dreamed of a layered world consisting of endless layers of chaos and order; a huge either energy accumulator, or its generator, or just memory cells.

8.8883. Information movement and conversion.

Movement is a source of energy, but energy is required for movement to exist.

The point is, you can't create energy in this world. The amount of energy in this world is the same and never changes. Movement is the process of energy conversion, the process of creating new information about its structure. Information moves and converts, changing the structure and form of energy. I see the meaning of such a system in the development and accumulation of information about matter, that is, about all possible forms of energy.

8.8884. Puff pastry theory.

Chaos movement is probabilistic in nature, but with the

increase in chaos, Probability is gradually turning into Certainty. That is, increasing in size, chaos becomes order. But as the system is puff, order, increasing even more, turns again into chaos. And that goes on indefinitely.

8.8885.

The system is developing dynamically. Fight is absolute, unity is relative. You are surely united with yourself, but your fight with yourself is endless. And so is all around. Any big substance consists of little substances and these substances are united, but endlessly fight with each other. Order consists of chaos, chaos - of order. The essence is that this fight is the source of movement and life. If you eliminate fight or unity, chaos or order, progress or reaction, everything will stop and die.

8.8886. Nesting doll.

Any big substance impersonates the order and form for little substances, forming its content in chaos. However, the same substance impersonates chaos relative to bigger objects, whose content it is.

8.942.4.

Any thing is a mass of things, some sort of mathematic bulk of essences. Thus, different people who look at the same thing, see it differently. Those who have better imagination, see more.

8.942.5. Polarisation.

In essence, it's possible to change the essence of things only by changing the viewpoint on them.

9.1049. Sun.

Chaos in chains of order, where chaos is content and order is its form... This system is not only beautiful, it is also the generator of energy, kind of the mini-Sun.

9.1050. Mind is the unity of chaos and order.

To gain access to energy, you, having become order, need to curb chaos. The stronger the order, the more energy it can squeeze out of chaos.

9.1056. It's little to know. You need to feel to understand.

There is no knowledge without love. Love allows you to feel knowledge, thereby understanding it. A person cannot feel what he is not interested in.

9.1103.

Thinking reasonably, there is a place for mystic in life.

9.1136. Confession is a tool to identify heresy.

In Romance languages the notions of «confessio » and «interrogation» are denoted by one word. During interrogation, a priest tried to find out if there was heresy in the soul of believer. Heresy is translated as «Objection»... that is, disagreement or some other understanding of religious dogma, different from the officially approved ones. Heretic can be put on the right track by various means. The essence of confession is to find out sins and vices and try to eliminate them.

9.1140.

The world is like boxes nested in one another. Within order there's chaos, within chaos there's order. Form represents order, contents represents chaos... This pyramid of generalities from the infinitely small to the infinitely big is the constant alternation of chaos and order.

9.1145. Pre-emptiness.

White colour is the full spectre, it's one. White is what aspires to black. White is the new beginning, the thing after which everything starts from the beginning again, it's the state of pre-emptiness.

9.1161.

Love is energy. Joy is energy. Money is energy. Material values are energy. Well, all of them are mutually convertible types of energy that exist in the body simultaneously...

9.1171. Beauty is a source of energy.

Beauty is a chaotic system in the chains of order, i.e. an energy generator. Beauty is energy.

9.1178.

Use is energy. Everything that energy can be called is use. Time is energy. Beauty is energy. Love is energy. Joy is energy. Uniqueness, novelty and originality are the amplifiers of energy.

9.1201.

Homosexuality is caused by inner contradictions. A child hears about a certain stereotype that a man should be this way, and a woman should be that way... And he feels too weak to match the pattern.... But if he doesn't match one pattern, maybe the second would be appropriate for him. The issue of self-identification is the issue of the choice of the pattern which fits you easier.

9.1215.

A reasonable person is the person who can defeat his emotions and passions. A human driven by his own passions can hardly be called reasonable.

9.1230.

The human's soul that does not think, does not reason, but

only feels is that unconscious IT, which Freud wrote about. The unconscious inner IT does not obey the Brain and intellect, but obeys the outer IT, social IT... Moreover, it is, apparently, some kind of alien nature, placed from the outside to control the person from without.

9.1250.

Harmony is when it is good here, and it is bad there, sadly, there is no other way. The system doesn't include possibility to take more joy than pain... There's less pain than joy.

9.1251.

The point is that harmony cannot be broken, but you can shift the balance point... Shifting balance into one or another direction does not disturb the WEIGHT of joy and pain, but only changes their duration in time.

For example, you can take great joy for a year and then suffer a slight pain for decades. Or you can take a huge pain for months and then live more or less normally for years. Balance shift in reality is extremely unpleasant and impractical. Addiction is when you rejoice one day and suffer seven days.

9.1252.

The more excellent a person is, the more fortitude he has, the bigger his carrot is and the more painful his stick is.

9.1254.

Harmony is when everything is 50/50, it's the easiest way to live. Other variants are more likely to be bullying than life.

9.1255.

Emotions consume a lot of energy, leaving almost nothing to the rest of the brain.

This is due to the fact that emotions emerged in those days when there was no rest of the brain, and they got used to acting independently. But now, when there is mind, logic and philosophy, it's time to finish with that.

9.1262. Maybe everything was the opposite.

Let's suppose that the human brain is a vivid example of symbiosis, when some parasitic organism, let's call it "the It", captured the brain of a mammalian monkey, thus provoking accelerated evolution of its mind, as an answer and counteraction to this parasite. Attempting to gain freedom and get rid of the parasite, the mind began to develop, gradually reaching the heights of the intellect and the fortitude.

9.1286.

Reasonableness is an regularity. The ability to observe regularity is reasonableness. The fact of existence of regularity in some system proves its rationality.

9.1295.

Hardships attack by pack. Wolves, good predators, rarely hunt alone, too. It's soothing that food also goes by herds... Good also loves company.

9.1301.

The essence of stoicism is in impassionate reacting to problems and not in being enthusiastic about getting into a mousetrap for cheese and carrot.

9.1313. Three types of logic in tridimensional space.

There are two types of logic, dialectic and quantum. However, there is also the third type of logic, figuratively speaking it's called Miracle.

9.1314.

From the point of view of evolution, it makes sense to reduce the level of emotionality in the organism of a modern human.

9.1316.

The idea of fatalism is created by the desire to absolve yourself of the responsibility for your actions. On the other hand, only people with fortitude can resist the fate.

9.1322.

Freedom is the ability to act according to your own will, being guided by your fortitude and your mind, not by emotions...

9.1323.

The key difference between pagans and Christianity is that Christians search for the causes of their pain and joy inside themselves, and pagans - outside.

9.1324.

The sense of fate and fatalism is that everything is subordinate to causal effects. It's impossible to avoid causal effects mechanism.

9.1331.

The light side of human soul is its form that has the content under it called its dark side. The form is chains of the content. In human souls chaos lives in the strong chains of order.

9.1335. An eternal enemy.

The struggle with emotions: laziness, passions, fears, boredom etc. should be seen as the struggle with the inner parasite. The inner symbiotic organism called IT craves for power but it's necessary to stop it.

9.1354. On the similarity of languages.

I've noticed this. My little child is bad at pronouncing letters. He pronounces the Russian word Voda (water) as Water (English «Water»). It makes me think that basically all the languages are notionally the same and the diversity of sounds is due to peculiarities of the sounds evolution.

9.1364.

The most terrible predator on the Earth is human, all the other filthy creatures will never compare to him.

9.1366.

Power is closely connected with opposition. If there's no opposition, there's no power.

9.1385. Internet minus 1.0».

To save the Internet from garbage, it's necessary to come back to its origins.

9.1591.

Human got on his feet, as he seemed bigger this way, and many predators started to be afraid of him and avoid him. The one who is bigger, is stronger. Human, crawling or running on his hands and knees has minimum chances of survival.

9.1592.

Human has no fangs, but he has hands. Predator needs hands and head more than everything else.

9.1620.

Harmonious love is the balance of energy exchange and energy conversion. Harmony is like the truth, as it's stable over time and can exist invariably for long time. Other kinds of love live rather a short life and destroy quickly.

9.1621. Principle of harmonic conversion of energy.

During balance of energy it's possible to convert one type of energy into another. The more types of energy keep the balance, the more stable such system is.

9.1629. Beware of those who don't love you.

There are only two substances in the world: energy and information. Everybody loves energy but those who love your information are valuable for you.

If a person loves your energy (money, appearance, body, love, joy, time, force and so on.), he has little value. Those who love your information (soul, thoughts, ideas, desires, abilities and skills) are valuable.

9.1653.

Mind is a creature of weakness rather than power. Mind is something that compensates for other types of power.

9.1671. Quantum observer.

At the microlevel, electron is only relatively orbiting atomic nucleus, in fact it is in every point of the orbite at the same time to a certain level of probability.

Human can be everywhere at the same time to a certain level of probability. All the possible variants of event happen at the same time, but the brain sees only one – the one it wants more, the one it's tuned to, the one it thinks of. The most likely, the brain builds the picture it sees itself, and then chooses physical analogue of it out of millions of accessible variants.

9.1672. Storm.

Usually all the problems happen at the same time, therefore

the system can't handle them and perishes... In this case, it's worthwhile to hide, avoiding the fight, shy away from clashes till the pressure of negative circumstances fades. It's necessary to buy time, hide and wait out the storm.

9.1707.

Pseudological riddles are the viruses that reprogram the sound logic of the brain... The person infected with pseudo-logic turns into an idiot.

9.1708.

Some riddles are like stupid questions in the sense that they can be absolutely dumb and far-fetched due to pseudo-logical constructions and reflecting upon them freezes and, thus, destroys the healthy brain.

9.1715. The fourth element.

The fourth type of truth is the truth beyond time and circumstances... The truth that has lost its motion, dead truth, absolute truth, truth of Nothing... The truth that exists always and never.

9.1728.

Progress is a consequence of contradictions and conflicts.

9.1753.

No, we are no titans. We are ants and our steps are small, even microscopic, so to say. We need to take a whole lot of steps to move and to live.

9.1782.

It's curious that the tasty and refreshing smell of cucumbers was made not for humans but for wasps to make them come and kill caterpillars.

9.1783.

Caffeine kills bugs and is slightly bracing humans. So, it's reasonable to suppose that it also kills those «bugs» that live inside a person's head.

9.1818. An empty place.

Truth is a strange power and should be hidden from those who have no love for it. On the other hand, only the chosen will be able to know truth, while others just will not see it.

9.1828.

Once you learn about the truth, do whatever it takes to possess it. Possession of the truth will make you rich and happy. The truth is light, light is energy and living base. However, it's only someone loving the truth who can possess it. Love is a necessary tool for getting the truth. Without love, you will simply rot of laziness, boredom and forlorness. Without love the truth won't seem interesting to you.

9.1830. The opposites.

A stick with two ends is very similar to a circle twisted in a spiral. Its beginning and end is one and the same thing, but at different levels.

9.1852. A dogma is the truth's skeleton.

A dogma is a dead truth that is no longer alive. A living truth is the one that changes with the circumstances that engendered it.

9.1869.

Why do you need to think if you are already clever? Clever people do not need to think, they are already clever ... Thinking is necessity for those who are foolish and weak, otherwise clever and strong people will simply eat them.

9.1875. There are not two, but three or four extremes.

There are four extremes in a four-dimensional reality and three of them in a tridimensional one.

9.1876.

A triangle is the truth, and a square is harmony.

9.1877.

The foundation of life are 4 chemical elements, and our world is four-dimensional, and it has four extremes, and harmony is four, and there are four temperaments, and even truth exists in four states simultaneously.

9.1878. Switching from one extreme to another is the easiest.

Character is what navigates temperament. The person with a strong character is able to switch his temperament from one state to another.

9.1880.

One and the same phenomenon can give an effect both direct and directly opposite, but you can choose the extreme that you like best by your willpower.

9.1886.

There are always three variants of extremes. By means of will it's possible to choose any of them.

9.1889.

Step by step one goes far. This is not a metaphor but a reality. If you do not rev the engine, fuel consumption is lesser, you get less tired and it's safer.

9.1901.

People are able to resist hypnosis if they know it is hypnosis, but if you are unaware of being hypnotized, it is difficult to resist it.

9.1902.

A decent person should know at least three languages: his mother language, a foreign language and the language of mathematics. Correspondingly, one's mother language is a written language, literature, ethics and philosophy. The language of mathematics is physics, geometry and algebra. A foreign language varies according to circumstances.

9.1913. Enemy to youself.

If people did not compare themselves with each other, there would be no greed, pride, envy and a heap of other sins... However, then man would compare himself with himself and become his own enemy...

9.1938.

A fool always sees knowledge only from one side, so he is unable to know the whole. The whole is the truth, for the whole vase is useful, but its shard is not. The shard of the truth is a lie.

9.1953.

Harmony loves Nothingness and Perfection despises Nothingness.

9.1990.

The truth needs lies very much, because mind is just what can distinguish the truth from a lie, to separate the wheat from the chaff. If there is no lie, there will be no mind, for it will lose its purpose.

9.2014.

Time does not pass. Time flows. Time is the river of life.

9.2015. Miracles should not be cited as evidence.

Do not cite as evidence something you saw on TV, on the Internet or heard from someone. Know that you are being deceived more than not deceived.

9.2051. A miracle is when all is Well.

Miracles exist. I've seen them hundreds and thousands of times and I've even learned to perform them... It is not complicated. All you need if Faith, Hope and Love and then all will be well. A miracle is when all is Well.

9.2053.

I do not see any crucial difference between the past, the present and the future. Formally, it is all the same.

9.2055.

People belong to ideas. People are the slaves of ideas. Slaves, however, cannot live without their masters.

9.2056.

Things are possible in different ways, both this way and that way.

9.2069.

This world is usually divided in two parts: darkness and light, truth and lie... It's as if the world was unidimensional... This world, however, is four-dimensional, so it would make sense to suppose that it should be divided in four parts (or three, depending on whether we believe or not in time).

9.2070.

I daresay this world is three-dimensional and time does not exist in it. Time is some persistent illusion related to movement. When there is movement, there is time; when there is no movement, there is no time. That is to say, time is very similar to light. Most likely, light is time.

9.2071.

Only an intelligent person can comprehend the profound meaning of the Bible, while a stupid one will only see a petty and stupid meaning in it.

9.2072.

If you want to learn something, you will have to learn everything on your own, since it is unlikely that anyone will want to tell you the truth. Everyone will stuff you with fairy tales, suggesting that you find meaning in them on your own.

9.2073.

If the brain is not switched on for a long time, it can become rust and get out of order.

9.2100.

If people did only what they could do they would not be able to do anything.

9.2131. If a person has no problems, he has no reason to think.

People do not like thinking most of all in their lives. If there were no problems and difficulties, people would have never learned to think and would have never got off the palm tree.

9.2152. A useful element.

Logically, it is fair to assume that God exists, but even if he does not, he should be invented for the good of the cause.

9.2153.

God is the strongest one around here. Well, nobody is stronger than him, so... we're just happy with that.

9.2174.

Nafs is the It. When Freud means the It, Muslims mean Nafs. Christians call the It a demon or the devil.

9.2175. Macaque (Rhesus) factor.

I do not think that the It (Nafs) is the soul. The soul is the human mind and intellect. The soul is the voice a person talks to himself with, this is a kind of social entity that lives in the cloud. The It is nothing more than an animal atavism.

9.2176. Limitless lying.

I found something that has no limit and this thing is lie.

9.2177.

The It likes pleasures and that is the only thing that the It likes. What is more, you have to feed the It, otherwise it will eat you. The question is what kind and how much pleasure should be given to it.

9.2181.

The loneliest person on Earth is God.

9.2182.

A person cannot understand what he does not agree with.

9.2189.

Wisdom and stupidity is not light and darkness. It is just a multitude of possible roads from darkness to light and back.

9.2191.

Anxiety comes when a person does not do what he must. When a person does what he must, he is calm. Rest is movement, lack of movement is anxiety.

9.2194.

Don't say Buddha. Say Buddhas. There are thousands and quadrillions of Buddhas. Buddha is he who exists in all of its forms simultaneously.

9.2196.

Goals are ephemeral: they come and go. Movement is eternal.

9.2197.

Moving only for the sake of a goal is trickery. Movement is inevitable and eternal, whereas goals are just road signs. Goals are made up to avoid boredom.

9.2200. The element of surprise.

There are thousands of ways to live. All you have to do is choose the one that suits you right now... Choosing is difficult and it would be much easier not to choose. Do not choose, then. Take any one at random and go ahead with it...

9.2207.

Human is a universal robot who can do what he can't do.

9.2208.

Human, unlike all other machines, can do what he can't do. Mind is the ability to do what is not their thing.

9.2211.

All processes need to be controlled. Uncontrolled processes slide into chaos very quickly.

9.2214.

You are a fish in the present, its skeleton in the past and you'll have been eaten in the future.

9.2218.

The thousand faces of Buddha are all those who have become a Buddha. You can become a Buddha too. The question is: What for?

9.2219.

In trying to get rid of an adversary, you just replace an old and known enemy with a new and unknown one. You'd better come to terms with your adversity, like it and make friends with it.

9.2221. The spiral of life.

Life is a never-ending process of creation and destruction on the road to perfection and back.

9.2223.

Everything can be very different at the same time.

9.2224.

You just have to stop searching and you will be found right away. The essence of this phenomenon is that what you have long searched for has become very used to you. Now that you are absent, it misses you. It is used to you and cannot live without you anymore.

9.2225.

Man is the only flaw of this world.

9.2229.

Why do you want to get rid of suffering? Do you no longer need happiness?

9.2230.

An impassive contemplation of one's own sufferings and joys is what stoicism is about.

9.2231.

The problem about the truth is not to achieve or to comprehend it, the problem is which truth to choose in each specific situation.

9.2241.

So much has been said about the harm of chatter that I'm telling you – chatter!... Talk drivel, pour all the foolishness out of yourself! You are the jugs overflowed by foolishness, you should pour it out of yourself to become empty and then fill yourself with wisdom.

9.2264.

Catching a deep person is harder than a superficial one. Superficial people live right by the water and can be caught with a fishing rod or a net.

9.2266. Quantum geometry.

Weird people sow one thing, hoping for another. What is ever weirder is that all kinds of things happen in our world, even this...

9.2268. Only if you come up with it yourself.

Impossible is impossible because it is impossible. Impossible circumstances are needed for impossible to become possible, but where will you get them from?

9.2269.

Before something new appears in the real world, it should be invented.

9.2280.

Much has been said about the fact that you can get what you want only by stopping to want it. The point of this idea is that the hunter should love hunting rather than the prey.

9.2281.

Being reborn into Faith, Hope becomes a reality. For hope to be reborn, it should first die.

9.2283.

These people do not bear any fruit, but they are beautiful. Beauty is a useful fruit too. This grass does not bear fruit, but it is useful, as cows that give milk eat it. That grass is not beautiful and nobody eats it, but it is also useful, as it makes you appreciate what you have.

9.2292.

When moving towards your dream, you follow the road from possible to probable.

9.2308.

A person's temperament turns him into a slave and a strong character frees him by tearing his chains off. If the character wins over the temperament, the person achieves freedom; if not, he remains a slave. At their birth, people are slaves of their wishes, but they can master them by developing their strength of character.

9.2399. Life is a game.

The war of the mind... An element of war... The intellective power and the emergence of éminences grises.

9.2413.

What is incomprehensible is a Miracle and what is comprehensible is science. They often ask whether Miracles exist. Just think, are there many incomprehensible things happening in life?

9.2432. The Armageddon.

If advertising disappears, trade will fall and millions of people will die of hunger en masse after losing their jobs. Economies will collapse, states will collapse and anarchy and war of all against all will start. 90% of the humankind will disappear and the survivors will understand that life without advertising is impossible and will resuscitate it, thus saving the human race from extinction.

9.2462.

Knowledge is the food for the mind. Brain needs food to think.

9.2465.

The more perfect the observer, the more he sees. The perfect observer, having even closed his eyes, sees a thousand times more than a fool with his eyes open.

9.2466.

Thinking style of men and women differs. Men prefer to think in solitude, while women do it in a team. Men are mostly interested in their own opinion, while women consider public opinion to be important. It happens because originally men were hunters and fishermen who had a lot of time to think in solitude, waiting for their prey, while women stayed in a big group of people doing the household chores and they had to think in a united collective mind without any opportunity to stay alone.

9.2483.

Extremes eat extremes. The rich eat the poor, the strong eat the weak, the intelligent eat the stupid and the haughty eat the humble... The more they eat, the bigger they become...

9.2491.

The more I was learning, the greater horror I felt at understanding how little I knew and would not have the time to learn because of the transience of human life and of my mind's weakness.

9.2504. The love of truth.

To learn truth, you have to have faith in it. Feel it to understand it.

9.2522.

Everything is possible and everything is impossible in this world. You can be successful in absolutely anything, while being unsuccessful in absolutely anything else. One may say that there are more probable events than improbable ones. That is an illusion, however...

9.2530.

God does not care whether you love him or not. God is truth. Those who love God can expect to get close to him and, thus, to obtain a little more additional life energy. Life needs energy and God is the information that manages energy. God is the strength that dominates energy. Love – the love of what it loves rather than the love of energy - is needed to manage energy. Energy loves perfection and loves those who love perfection, who create perfection, who serve it and who work towards it. Energy loves information. God is perfect information. God is truth. Information, however, is the God of this world only. Above it, there is that very NOBODY who created and invented it all.

9.2535. Origin of a reasonable human.

Initially, there was only the It in human, it only ate, sought pleasure, bred and slept. Then Love came. Love made the It desire something more than simple pleasures. Love showed the It that there is a stronger pleasure and this is THE TRUTH, this is INFORMATION... Love managed to defeat

Laziness and force the It to work and learn, improve and grow... Thus Mind came to life.

9.2536.

Any person can call himself God because God lives in everyone. There is nothing special about the fact that God lives in someone. We are happy for this good person. Those in whom God lives are happy and joyful; they are happy and satisfied with their lives. We feel more sorry for those, in whom God does not live: these people are weak and unhappy. The shrieks of the godless suffering in Hell give our hearts pain... But how can we help them? They reject love, they reject truth, they reject God... There is nothing we can do to help them.

9.2537. Аккумулятор энергии.

Тот, кто, лежа на солнышке, наполнился солнцем, как аккумулятор электричеством, не становится от этого ни солнцем, ни генератором энергии. Однако, конечно, со стороны взгляду неопытному может показаться, что подобный аккумулятор есть некий эталонный светлый и чистый источник энергии, полный добра и света. Увы, сей источник энергии вторичен и требует постоянной перезарядки. Хотя, конечно, аккумуляторы полезны.

9.2537.

He who, lying in the sun, is filled with sunlight like an accumulator filling with energy becomes neither the sun nor an energy generator. Of course, it may seem to some untrained eye that such an accumulator is some reference clear and pure source of energy, full of good and light. Alas, this source of energy is secondary and requires constant charging. Still, accumulators are useful, for sure.

9.2538.

A judgment that is most suitable for a given moment is truth.

9.2541.

Truth is when something is whole and lie is a piece. The out-of-context truth also turns into a lie.

9.2544.

Given a free rein, evil increases, first, twofold, then three-fold, then tenfold and then dies of hunger, having eaten up everything within its reach. It was so before, though. Now evil grows food for itself and can grow eternally.

9.2581.

I have found that human souls are rarely older than thirty, and they grow even younger over time. It is probably due to the fact that human lifespan was 25 to 30 years old for millions of years and the human soul does not know that living longer is possible.

9.2607.

It is not money that rules the world – it is beauty that rules the world. People want money to make their lives more beautiful.

9.2613. NOT your business.

Doing something that is not your business is very dangerous. If you are not doing your business, you take someone else's place and rob people of their meaning of life. People who are not doing their business get beaten and scolded, adversities fall on their heads, they get sick and suffer a lot.

9.2633.

Age is about how old the soul, not the body, is. The body, however, influences the soul, so pain and a lack of energy make it older.

9.2638. Understanding obvious things is most difficult.

Obvious things are the most improbable of all. Maybe that is why many people avoid obvious things and try to complicate everything.

9.2642.

There is Nothingness, there is Perfection and there are roads leading to them. There is the true road, there is the road of lies and illusions and there are mixed roads. If you love your goal, you will get to it, no matter which road you take.

9.2651.

Traveling in time is impossible because time does not exist, but speeding up or slowing down in it is possible. This is almost the same thing.

9.2657. The dial of eternity.

Time is a spiral, not a circle. Time passes along a spiral.

9.2667.

A person may be talking deliriously... He is talking and talking and you are sluggishly listening and listening while reflecting on something else and then bang!.. You come up with a beautiful, nice, captivating, fantastic and extremely useful thought...

9.2668. Seeds of life.

The Universe emerged from a grain, rather than a simple point. This grain germinated, grew and yielded... One grain gave birth to thousands and millions of new, small universes.

9.2669. Diversity of universes.

Observations of nature show that seeds of life are different and even the universe itself may emerge from a grain and it may equally emerge from beans...

9.2672.

The magic of prime numbers is powerful and their nature is clear. There is something weird about 6 and 9.
6 serves 7 (seven is expanded one).
9 serves 8 (eight is the harmony of 0 and 1).
Interestingly, these two digits can easily change their masters and turn into one another.

9.2673.

As a matter of fact, one can serve both harmony and perfection. Perfection is seven and harmony is eight. Six serves perfection and nine serves harmony. Six and Nine are the two extremes that easily turn into one another.

9.2710. Unhealthy egoism and avidity.

The Devil is he who loves God most of all. The love of the Devil is extremely jealous as he wants to have God all to himself... and he hates, out of jealousy, all others who love God... and he hates, out of love, all those who do not love God.

9.2712. The integral picture of Existence.

In creating Variothoughts, I wanted to create Truth. Truth lives eternally. Nothing can destroy truth – neither man nor time. Truth is a whole. Parts of a whole, pieces of truth are nothing more than lie. Lie dies quickly; lie is very afraid of time. I devoted a great deal of effort and energy to create an integral picture, to create a whole.

9.2718.

When tired, a person feels much more intelligent than he is

in reality.

9.2719.

In their youth, people feel much older than they are, but they grow younger with time, stop growing at the age of about 25 and remain so until their death...

9.2722.

Noise prevents people from reasoning. The louder the noise, the less and the worse people reason. Noise turns their brains off. Only in silence can they turn their brains back on...

9.2724.

Grains live in circles and plants live in circles... Rational people live along a spiral. Frankly, everyone is living along a spiral, but the angle is always different. The more rational a person is, the greater the angle.

9.2727.

The world is a game and a good game too. Everyone plays by the rules in a good game. This is why any talk of injustice is the work of the Devil... Nobody can break the rules of this game, not even God. By the way, the Devil also must play by the rules. Who knows what about the real rules is another question. Truth is the real rules of the game named Life. Lie is the false rules. Breaking the real rules is impossible, whereas breaking the false rules is possible.

9.2728. Price of a spiral.

All human problems are a consequence of the violation of rules, but the very opportunity to violate the rules is provided by the system and also occurs according to certain rules and is accompanied by problems and suffering. Why is this done this way? - The system becomes more stable, flexible and capable of evolution this way.

9.2729. Slightly.

The opportunity to slightly violate rules is pre-planned by the system for the sake of its stability and flexibility. However, the more severe violation of rules is, the more probable the negative reaction of the system is. Breaking rules should be done quickly and slightly, otherwise emergency safety systems will spring to action.

9.2730. Rubber rules.

Slight violation of rules is a result of the system's flexibility rather than mistake. Schedules should be flexible and not stretched over circumstances. Stone schedules are not suitable for life. Life's more rubber than stone.

9.2734.

Life is an absurd question, which has an infinitely large number of answers.

9.2737.

The evolutionary cycle of civilization from creation to extinction is 8,000 years, and then it moves on to the next cycle and to the new stage of existence.

9.2739.

A medley of peoples is a political situation in the early 21st century that has to do with the intention of European political elites to refresh European blood by following the example of the USA that brought the best people from Africa, the Near East and Eastern Europe.

9.2746.

If you have a vice that you want to kill, overdo with it to the max. I once played a game for three weeks in a row and I hate games ever since then.

9.2748. There are at least three explanations for everything.

9.2757. Up and down.

The even turning of millions of gearwheels and wheels pushes the world up the road twisted in a spiral. It will never turn off and it will never stop. Everything is designed in such a way that something compensates for something else. You will ask me what good is. My answer is: good is balance. You will ask me what evil is. My answer is: there is No evil. Evil is impossible. Imbalance could be evil, but disturbing balance is impossible. The wheel of life will turn eternally as long as some will push it upwards and others downwards. Those who stop die.

9.2758. In the wrong place.

Life is truly dangerous for those who are afraid of it. Being afraid is very dangerous. Fear implies that one is weak, one is not ready, one is not doing his business and is in the wrong place. A person not doing his business outrages existence. Everything literally jumps, teeth bared, at this poor fellow in an attempt to take him out of that wrong place. The poor fellow will be saved only by finding his right place.

9.2760. Grain

Everyone can think big thoughts. Big thoughts are clearly visible... But I like small thoughts. My thought is very small, like the point which this world came into being from.

9.2764.

Unnecessary thoughts clog the head like rubbish clogs clean water. Then, after the rubbish settled to the bottom, I saw the truth in the crystal clear water ... "Wow," I thought, "we just had to remove the unnecessary."

9.2766.

To leran the truth, you need to remove the unnecessary.

9.2767.

The easiest way to kill the truth is to fill in it with rubbish. Under a thick layer of mud, The truth forgotten under the thick layer of rubbish can be preserved infinitely.

9.2775. The heavenly garden.

What is it you do not like about this world? The world is just fine! Sin less, work more and reason more often... and every-thing will be great.

9.2778.

The danger of trying to extract a meaning from something meaningless is that it is possible to extract any meaning from meaningless things and this freedom depraves the ob-server a lot.

9.2784. Nothing is ever enough for a creator.

A creator is a person for whom nothing is ever big enough. Nothing is ever enough for a creator. It is because the creator is always creating something. Having created something, he starts creating something else, something even bigger and greater. A creator creates a new world and we remember from experience that the creation of the world is a process stretched into eternity.

9.2793.

Many people do not understand words, because it seems to them that words consist of letters, but this is a terrible delu-sion, words consist of sense.

9.2797.

Grand concepts and secrets of existence. We are reading The Theory of Existence and enjoying the show.

9.2821. Замкнутая энергетическая система.

Человек самовлюблённый это зацикленный человек. Он любит сам себя, сам себе отвечая взаимностью. Любить такого человека со стороны не хочется, но часто мы любим плоды его трудов, а с ними вместе и их хозяина.

9.2821.

A self-enamored person is an obsessed person. He loves himself and loves himself back. We prefer not to love such a person on the outside, but we love the fruit of his labor and their owner too.

9.2822. A person in love with his own god.

One does not feel like loving self-enamored people, but they are sometimes so good that there is no way to resist attraction. Powered by themselves, such people can be extremely strong.

9.2869.

The main problem with any wish is that it is not really worth it... If it is, however, hardly anything will keep a person from carrying it out.

9.2972.

Money is just energy and it serves Information, just like any kind of energy. What information is that? – It is commonly called Truth. Money serves those who possess Truth. – Who possesses Truth? – Those who love Truth. Love engenders faith and faith gives them POWER. Power and truth allow them to control energy, matter and money.

9.3025.

A half of verity is a lie. But one-and-a-half verity is a lie too.

9.3029.

Jihad in the Way of Allah is a road to achieve Truth. Allah is truth and, on his way to it, man should eradicate sins, vices and lies from his soul...

9.3040.

People have various names for God. I have heard the following of his names: Truth, Love, Providence, Destiny and the Blessed Star.

9.3059.

There is no rule that cannot be violated if the situation so requires.

9.3072.

Rules are what you get beaten for when you violate them. People often forget rules or are even completely unaware of them, but anyone can learn them the hard way. If they do not feel anything for violating rules, the latter are no rules at all but total crap.

9.3073.

I saw a fish, whose DNA contained nothing unnecessary, but human is full of the unnecessary, that is probably why he is human, not fish.

9.3080. A teenager.

Considering that the age of the souls of all adults are maintained at about 25 years, teenage years last until the age of 24.

9.3090.

A world can be built from anything, even from Nothingness.

I have seen the most beautiful worlds made with nails, screws, gearwheels, spark plugs and bike chains... wide sand and stones, shells and dry twigs...

9.3091.

This world consists of straight and curved lines... even light is not as straight as it seems to be, but rather curved and undulating.

9.3092.

Straight lines are all alike, like twins, and bent lines are all different.

9.3093.

People are like nails, each one is crooked in a special kind of way and none is straight.

9.3096. Information is distorted, passing through the atmosphere of the human soul.

The human soul is a kind of atmosphere, whose composition is always different and varies from person to person. Therefore, it refracts the solar white rays in different ways each time. The gas component of the human soul is different everywhere, so the refraction of colors always occurs in a special way. The color of the sky and the colors around are always different.

9.3103.

Choose any of two extremes – the consequences will be the same.

9.3104.

Man is an electrically controlled hydraulic system.

9.3105.

Many talk of the fear of death, but few talk of that of life. Let me remind you that the main objective of torturers was to prevent prisoners from accidentally dying too early.

9.3108.

The point of religion is that not only the chosen and the fortunate, but also the intelligent should have control over energy. They say God can choose a man, but a man can choose God too. A man can become the one loved by God – he can become the chosen one.

9.3109.

Masons were not the first - Jesuits, Knights Templar, Sufis and even early Christians preceded them... The point is that the intelligent should have control over power, rather than those who happened to be lucky.

9.3111. A teleporter.

Teleportation is, essentially, the technology of eternal life. When assembling a body in a new location, one can assemble it the way one wants... One can assemble a new body, modify it or, if desired, one can even turn into someone.

9.3112. Riding a broomstick.

Air is a discharged liquid. Essentially, it is water and its diamagnetic properties make it possible to swim... to levitate in it. All that contains water can navigate in air, i.e. levitate. In particular, 80% of the human body and of plants is water. That is, broomstick riding and levitating magicians are possible in terms of physics.

9.3114. Graphene and the paradox of the heap.

Something big consists of something small, but the latter does not exist. It exists in a heap, but, in isolation, it does not.

9.3115. Multiple functions.

Many things cannot exist, but they do exist, especially if they are many. Multiplicity gives them the strength to exist as objects.

9.3116.

Winning over oneself (philosophy), victory of reason over the Id (according to Freud), victory of the spirit of Nafs (Islam), victory of reason over emotions, victory of one's idleness, vices and so on – those are all the same thing...

9.3125. Twenty-one.

Too early is as bad as too late. Everything should be in time. Unripe fruit and overripe fruit are equally useless and inedible.

9.3131. Shutting down the brain.

A great way of shutting down the brain is used in meditation, rituals, hypnosis, zombification and so on.

Repeating uniform or chaotic movements that do not require the brain to function, yet consume the energy of the body, shut down the brain. That is, all kinds of dances, chaotic and systematic body movements, systematic and repeating movements, imitation of others, aerobics, fitness and yoga... shut down the supreme nervous system (reason), leaving only low animal instincts (the It, according to Freud). Being a primitive man for a while is a narcotic pleasure. People also like shutting down their thoughts, the source of pain. This has a number of dangers, though. A man can be easily reprogrammed and hypnotized in this state, and it is also a highly intense, habit-forming pleasure. Another great danger is that the Animal essence will win over and enslave reason.

9.3150.

A truth that always exists never exists. It exists beyond time and space, beyond all circumstances. It does not exist, however, in the real world of situations in which there are no circumstances, time or space. A truth that is independent of circumstances is an abstraction whose existence is impossible in the real world.

9.3156.

Can differently dimensional entities exist within the same universe?

9.3157.

The fact that minus times minus equals plus proves that extremes are easily interchangeable.

9.3158.

The fact that minus times minus equals plus proves many paradoxes of the real world. Extremes can easily change places. Extremes are interchangeable. Extremes engender extremes. Take a step beyond the edge and you will get to the beginning. Truth engenders lie and lie engenders truth.

9.3159. The interpretation of simple notions.

The problem with many people is that they know words without understanding their meaning. Few know what Love, God, Truth, Money, Joy, Beauty, Uniqueness, Chaos, Order, the Damned or Angels are. In fact, only energy and information, which controls it, exist in this world. Matter is just information about the form of energy.

Love means serving one's work, cause, ideas, truth, objectives, people and, generally, the world; and, last, it is about sex, boys and girls. In fact, love is a kind of energy easily convertible into other kinds of energy. Love is the most powerful and easily accessible source of energy. The signs of love are kindness, respect, mutual trust, joy, attention, interest, politeness, absence of avidity, absence of idleness and disregard for flaws.

God is truth, providence, nature, the order of things, fate, the blessed star and so on.

Truth is the information that controls and structures energy (matter, objects, money, love, etc.). The more truth is perfect, the more it has access to energy. If truth is false, it is called Lie, and energy does not obey it. The notion of uniqueness is closely related to the perfection of truth, for identical perfections kill each other. Truth is the function of circumstances; and some specific truth corresponds to all unique circumstances. In a three-dimensional space, three variations of truth exist simultaneously at one moment of time. Lie is what does not correspond to circumstances. Lie is half the truth or one truth and a half.

Money is one of the main forms of energy. Its major advantage is its easy convertibility into other kinds of energy. Money is fuel, water and benzene. There is no life, growth or creation without money.

Joy is one of the forms of love (energy) as well as a signal for readiness to love, i.e. to participate in an energy exchange. Love should always be responded to with joy. The absence of joy shows that the person does not want to be loved and does not want to love back, that is, he does not want to participate in energy exchange operations. To sum up, if you need love, rejoice, and if you need money, rejoice too.

Beauty is a source of energy, joy, good spirits, luck and so on. Beauty is an energy accumulator and something that has achieved perfection and is now able to control, accumulate and distribute energy. It is able to transform light into other kinds of energy... Beauty is a balance between chaos and order, where order has structured chaos and chaos is a source of uniqueness and energy. Beauty is a nuclear reactor, in which chaos structured by order generates pure energy.

Chaos and Order are form and content, zero and one, a source of movement, a source of energy, a source of life and an energy generator. Chaos is a source of novelty, freedom and energy. Order is Benefit in the sense that it structures chaos for the benefit of the cause. Essentially chaos is energy and order is information. Harmony is a balance between chaos and order, when these two entities operate as an energy generator.

The Damned are the ones who worship perfection and reject Balance. Angels are those who also worship perfection, but they consider that it needs a Balance with NOTHINGNESS... In other words, Angels serve harmony and think there should be a balance between good and evil, black and white, whereas the Damned think one entity – one – is enough. The number of the Damned is 7 and they are served by 6 (six). The number of the Angels is 8 and they are served by 9 (nine). 6 and 9 are people deciding whom they will serve and, interestingly, 6 and 9 easily change places, turning into one another. Angels rule over the carrot and the Damned over the stick.

9.3163. It is said order will be born from chaos.

Forests are, first, planted and, then, thinned out. That is, order is born from chaos.

9.3170. A good motive.

Advancing by the path of most resistance is more interesting...

9.3172.

A queue is a way to avoid hold-ups. A queue avoids friction, which decreases energy and time loss.

9.3173.

Man is a pet: he feels hungry and cold at liberty.

9.3185.

There is no time in absolute darkness because there is no movement or light.

9.3186.

In darkness, time drags slowly. It slows down but does not disappear because there is, after all, some light...

9.3187.

To reason, it is necessary to liberate reason. However, after coming up with an idea, reason should be put back in chains in order to put it into practice, otherwise the body just refuses to obey.

9.3190.

The brain is a machine, if it is not used for a long time, it can get out of order.

9.3198.

What seems truth at first glance may prove to be an outright lie at second glance, but you should wait for the third glance before making a judgment.

9.3199. Clarification of the Pareto principle. Soloinc's 13th law.

Theoretically, the Pareto principle about 20% of efforts is wrong, as for four-bit or eight-bit systems it's more logical to talk about 12,5% - 25% of efforts for the maximum optimal coefficient of efficiency of one process to 70-80%, and the amount of processes from 3 to 7. Thus, 25% of efforts give 80% of income, which goes in 3 directions to give the maximum coefficient of efficiency of 240% or 12,5 % for 70% of success, and 7 possible directions, with maximum coefficient of efficiency of 490%.

9.3210. Disintegration of time into energy.

The volume of energy in the universe is constantly growing due to the disintegration of time into energy.

9.3211.

The beginning of a circle is a point. The circle is born from a point.

9.3212.

Common sense is the ability to get out of a point of view and build a three-dimensional pattern of reality that can be seen from all its parts, both inside and out.

9.3213.

Where an idiot sees a point, the clever one sees the whole universe.

9.3214.

The Universe is a four-dimensional dot within eight-dimensional space.

9.3219. Social neutralizers.

The main purpose of school, universities and army nowadays is not to teach people life or prepare them for something, but more likely it narrows down to getting teenagers occupied by something and moving them away from the streets. However, computer games are great at the same task. Teenagers in the streets are a disaster for the state.

9.3266.

Harmony is not a point. Harmony is a three-dimensional sphere that combines all extremes simultaneously. A point cannot combine the incompatible. To do so, it should grow into a three-dimensional sphere.

9.3267. Any point is an endpoint on a straight line.

Any point is an endpoint in a unidimensional space, and in a bi-dimensional space too...

9.3274. The algorithm of energy packing.

A space of any dimension is mutually transformable and able to fold in and out one another using these or those mathematical algorithms known as truth. Each space dimension has its own algorithm of energy packing: the lesser the space dimension, the greater energy concentration.

In a unidimensional and three-dimensional object, for example, in a unidimensional point and a three-dimensional universe there is an equal amount of energy, but it has been packed using different algorithms.

9.3275.

Any space is always simultaneously unidimensional and endlessly dimensional. A space is simultaneously uni-, bi-, three-, four-, five-... seven- and so on dimensional.

9.3280.

There is no uni- or three-dimensional space; there is always time. Space does not exist in the real world beyond time; it is, however, possible theoretically.

9.3281. Time is one.

Time is not a measurement. Time is nothing. It is that very zero that is necessary for the system to advance. As long as the system grows, it has balance in it; when movement disappears, time disappears too. If there is movement, it is one; no movement is zero. Inside itself, time is divided into uni-, three- and seven-dimensional spaces. The movement of time is the complication of one by three, seven, twenty-one and so on.

9.3289. Revisiting the energy conversion efficiency.

The energy conversion efficiency is 100% true only for the transformation of energy inside Newtonian space from one kind into another. In case of processes of energy packing and unpacking that take place between one- and three-dimensional spaces, there can be any whatsoever energy conversion efficiency. Inside a onedimensional point, there is as much energy as inside an entire three-dimensional universe.

9.3290. The evolution of the universe.

The universe evolves from a unidimensional to a multidimensional space by storing information. The exponential growth of information leads to a collapse of the mass under its own weight and to the system's production of critical mass and a supernova explosion. Specifically, there are two explosions: one comes outside and the other inside itself. The explosion inside itself produces a breach into Nothingness, a superconducting point, on the other end of which a new universe emerges, of greater dimensionality than the current one. Energy unpacks and flows from this three-dimensional world into a new seven-dimensional universe, thus creating it. The supernova gradually works off its energy in this world, becomes depleted and dies out, giving less and less energy to the new world. When energy is up, the new world, bereft of its source of energy, will collapse under its own weight and, folding back into a point, will produce critical mass, and everything will start over again.

9.3291. Who created the Universe?

There is no time in a one-dimensional space, that's why the system can create itself on its own.

9.3294. Overreach is bad. Best is the enemy of good.

The notion of balance can be shown on reaching the critical mass in the nuclear reactor. If there is too little fuel, it's underreach and nothing happens, if everything goes moderately, the reactor generates energy and if it reaches the critical mass, there's overreach, then there will be a nuclear explosion and the system will be destroyed.

9.3326. Plus the time.

Foolishness in one-dimensional, mind is able to observe the development dynamics of three-dimensional reality. The maximum that foolishness is capable of is to become a flatness. Flat two-dimensionality allows foolishness to consider the behavior of a point in dynamics.

9.3332.

Everything strives to help the perfect and the big, making it more perfect and stronger.. The small, wanting to get energy from the big, always aspires to help or serve it somehow, find its imperfection and offer to make it better. The small makes the big greater, the big is just a lot of small. The greater and stronger the big is, the more of those who want to serve it attracts. There is much energy near the big and strong.

9.3341.

Faith is the most effective weapon. Faith works wonders. The person believing in himself is capable of much.

9.3342.

All people are approximately the same and those who differ are not people at all, but somebody out of books and TV, somebody artificially created in the depths of the Internet or someone's sick fantasy.

9.3352. The fourth state.

This world is three-dimensional, there is past, present and future. The fourth state of time is no time.

9.3370.

Human is a hi-tech means of production, it's dangerous to invest in it if you can't use it. People are different inside: there are more or less complicated ones. The smarter, more complicated and educated a person is, the harder it is to use him. Nevertheless, the more hi-tech the means of production is, the more profit it brings, the more surplus value it can create.

9.3371. Business is always a system.

Business is a multidimensional business- blueprint that exists in moving fashion. That's why running business demands brains and the ability to see the tridimensional dynamic picture of reality. It's hard for a stupid narrow-minded person to be successful in business. Monodimensional and two-dimensional flat people are barely meant for business, business demands a four-dimensional view of life.

9.3372.

Business means transformation of time into money. Time is the only resource that is almost equally distributed between people.

9.3453.

Observing time for a long time, I've noticed that time is cyclical...All these early, in time and late are repeated endlessly, inevitably replacing each other, disappearing and appearing again.

9.3554.

Pieces of the truth are a lie, but if all the lie is united, it will become the truth.

9.3571.

Fear is a chemical reaction that occurs due to the release of hormones in the brain. Fear paralyzes the resistance of weaker individuals, it's a submission hormone. Evil and strong individuals cause the weaker individuals' desire to obey them. The excess of the testosterone hormone is characteristic of dominant individuals and makes them especially evil. Curiously, the weaker individual obeys from fear and the stronger one attacks.

9.3572. Zero and One.

"I don't want" and laziness are the normal state of the human soul, the state of Nothing, the state of Zero. To switch, the brain needs to generate hormones, for the release of which such brain signals as love, hunger, fear, pain, etc. are responsible. It means that to go into the state of One (movement), some motivational signals or effort are needed.

9.3575.

The human brain has a built-in leader submission program. If the brain determines that the leader is strong enough to provide the community with food, energy and safety, this program turns on and makes the weaker individuals obey.

9.3576. Built-in programs.

All the programs built into the brain are in the human unconscious It and are almost beyond the Mind's control. To deactivate programs, the IT should be deactivated almost entirely. All programs work at the level of feelings: I want to - I do not want to, I like it - I do not like it, I feel lazy – I do not feel lazy. The most famous programs include: the leader sub-mission program, the fear program, the love program, the "you are food" program, the self-destruction program, evaluation programs on dominant individuals, etc.

9.3704.

The longer a person lives, the faster his time flies. But God lives forever, his time flies superfast.

9.3705. Time is one-dimensional.

9.3706.

The more you do, the more your life is filled with tasks, the slower your time goes, so the more time you have.

9.3720. Reaching of the truth.

If pieces of the whole are put together, it won't become the whole. To make the whole, one should totally get rid of these pieces either by merging or by gluing them.

9.3721. Synergistic effect of integrity

The whole is what has synergistic effect, emerging as a result of uniting its parts. If parts are put together, they will never make the whole, as the whole is always substantively bigger.

than its parts. You won't get the whole just by summing the parts. 2+2 = 22.

9.3722.

It is interesting to note that the system properties are always different from the properties of its components' parts. That is, by adding different entities to integral systems, you can get previously completely unknown and unique properties.

9.3723.

Form is what organizes essence, giving it integrity. Essence is like chaos and consists of thousands of disintegrated pieces. Form creates integrity by uniting these pieces of essence.

9.3724. Properties of a mask.

Each form has these or those properties. Choosing the form (mask), you automatically accept all its properties.

9.3727.

Human behavior is a balance of inner desires and external restrictions. Human consciousness is subjective and human behavior is objective. The person inside and the person outside himself are completely different people, and the one who is outside is the mind while the one who is inside is the unconscious It. The task of the mind is to control the It and not to let it begin to control the external behavior.

9.3728. Property of the mind.

The stimulus causes reaction, that's true. Nevertheless, the main property of the mind is to distort reality, therefore one and the same stimulus can cause different reactions in different people. Of course, the simpler the stimulus, the more predictable the reaction is... But even simple stimuli, such as a stick and a carrot, always have three equal reaction options.

Deception is the defense of the mind from the stimulus. The mind can formally submit to the stimulus, continuing to seek the opportunity to deceive it and act differently from the expectations on the reaction.

9.3729.

The more stupid a person is, the easier it is to control him through stimuli and reaction. Mind tends to deceive stimuli.

9.3730. The first human need is the thirst for joy.

Three primary human emotions are love, anger and fear...
Mingling with each other, they generate seven more.

9.3731. Pseudo-random code sequence.

The trial-and-error method is, in fact, the "stick and carrot
method". A person tests what is good and what is bad by
touch, forming his own world view. Moreover, the more experiments he conducts, the more complete world view he
builds. In fact, this is the personality coding method, a sequence of zeros and ones coming from the outside world and
forming a human personality.

9.3733.

The body is inseparable from the soul, but it is separable
from the mind. The soul is the product of the body and its
desires, while the mind is the product of the outside world.

9.3734. Common cause.

Classically, soul and mind resist each other, mind tries to
subordinate and control soul, soul tries to do the same.
However, it's worthwhile to talk about symbiosis of these
two essences and uniting them as a whole to achieve synergetic multiplying effect. Mind, being the form, should control soul - the content, channeling its desires.

9.3736.

Cognitive psychology considers psychic to be a device
with fixed ability to transform signals, but... this approach
doesn't take into account the fact that human has mind.
Mind is the external supersystem of human psychic, changing the logic of work from the fixed to the pseudorandom
one. Mind decodes external signals by some random way, dependant on its internal settings. And even the very psychic is
able to work in totally different modes, switching between
them from time to time.

9.3738.

The human brain is a computer system that implements the algorithms of pseudorandom logic. Having added two plus two, such a computer system can give any result, which, nevertheless, will be justified by some well-founded logic, formed randomly.

9.3756.

The problem of self-study and reading books is that the understanding of the material depends on the level of perfection of the reader, his mood, his interest in the material... In other words, without a living mentor who would sit and control it all, this process is ineffective. Moreover, almost all knowledge works only in the system, and people tend to think in dots.

9.3758.

Fake love is love for your own pleasure. Real love loves its object, it's very strong and nothing can break or stop its desire. And fake love is very easy to break, you should just take away its pleasure. It will die from one word, one stumbling rock, one offence and so on.

9.3768. The one who fears loses.

Fear exhausts forces. Fear turns on the «you are food» biologic program. «Freeze and don't move, you are about to be eaten. – Nature tells you – The weak is food of the strong, good predators have come to eat you, don't resist»...

9.3769.

Obstacles are necessary for overcoming them joyfully. Without obstacles many reasons for joy will disappear.

9.3799.

When normal behavior doesn't bring them joy, they become criminals. A criminal is the person that gets pleasure from asocial and abnormal behavior.

9.3800. Real crime.

Capital punishment is a salvation and mercy for many criminals, their yearning for self-destruction makes them happy. A suicide is happy, but he can be punished with a terrifying and very long life.

9.3817. Supereffort.

Many people talk about some superefficiency, reason upon giving one hundred percent. The truth is that if you give at least 20% and switch on your brain, it will already be "bingo" and becoming 1% of successful people...

9.3831.

The Bible is all about metaphors and parables; don't take it too literally or pick on every word.

9.3832. The price?

Whether there will be demand for an item or not, depends on whether it meets some human needs from Maslow's hierarchy of needs.

9.3833. Children's joys are multiplied parents' joys

Everything that concerns children in Maslow's hierarchy is in first group of needs and is related to sex... In particular, children raising, education and their achievements belong to this group.

9.3839.

Civilization allows the weak to survive and weakness is the basis of mind. The weak, trying to achieve perfection at least in anything, reaches truly unique heights.

9.3841. In the beginning everybody was propagating by germination.

Who was the first - man or woman? Egg or hen?

9.3868.

To avoid going crazy from eternity, the immortals speed their time up, their minutes are our years. It's this phenomenon that explains why God reacts so slowly to the people's prayers and sins.

9.3875. Personality code.

The child encodes his personality through the trial-and-error method. He can get one or zero, satisfaction and joy, pain and nothing. Actions that give satisfaction and joy become positive patterns and those that bring negativism are rejected.

9.3876.

Space is matter, too, just the discharged one.

9.3891. Imagy-therapy.

The achievement of soul-mind symbiosis by comprehension of the subconscious. The achievement of the synergetic wholeness opens inner soul's sources of energy, inspirits and gives vigour and good mood.

9.3908.

The essence of Imagy-therapy is picturing of hypothetical situations from the past or the future. The analysis of possible scenario and events from absurd to quite real ones. On the basis of pictured situations the analysis of real situations is performed. Imagined situations are used as some kind of training. It also helps to find out logical and philosophical mistakes, negative automations and triggers.

Where a person is mistaken, why this person is mistaken and what should be done to avoid mistakes.

9.3919. The fourth state – current process (now).

There are four times: past, future, present (the processes that are not over yet and keep affecting now) and the main process - now.

9.3927.

Circle is zero. Everything started from Nothing. Point emerged first. Malevich's Black Square emerged much later, when circle had already died and there were very many points.

9.3928. Imagy practices.

Imagian-practice, Imaginc-practice, Imagency-practice, Imagenial-practice, ImaGym- practice.

9.3929.

Knowledge is what is already realized and what is not yet realized is black matter.

9.3930.

Substantia nigra is the pre-state of substance that is about to become matter yet.

9.3931.

Circle is the evolution of point, the next step of evolution is blast. Splitting, circle forms various geometric figures.

9.3935.

The fourth state of time is when there's none of it. This very momentary process (now) when there's no time but decisions about the direction of motion should be taken momentarily quickly. In essence, we live in this supertemporal moment when there's no time but there is motion.

9.3955.

«Imagy» therapy works in the forth time called "Now" (the running process). Now all decisions are made by means of a person's "EGO" (the mechanism responsible for making decisions). EGO is influenced by mind (intellect, logic, wisdom, patterns) and the subconscious (soul, it, feelings, automations). The subconscious and mind are influenced by things and events from the past, the present and the future. «Imagy» practice places special emphasis on formation of the future (dreams, visions, hopes). The future rules the process "Now" very effectively. The past is considered by means of psychoanalytic approach (childhood, dreams, growing-up and adolescence). The present is analyzed with the use of gestalt therapy (here and now). The techniques of cognitive psychology are actively used. The main principles of Imagy include the comprehension of the subconscious, recognition of delusions, analysis of causes and consequences, tension and stress reduction (the sources of energy loss in the body), the clearing of brain from rubbish and logical errors, the correction of negative patterns and automations, the control of emotions and desires, the obtainment of the symbiosis between mind and the subconscious in order to improve efficiency.

9.3956. Different view of extremities.

Imagine our life is an Earth-like waterball and we live on its very edge, on border between it and Nothing. We, people, are edge creatures. We have to dive deeper to get energy or wisdom. The deeper, the more energy there is. The deep truths are the most valuable. But energy is also somewhere in Nothing, not only in the depth. In short, the most foolish and poor in this world are shallow people full of extremities, and those who dive deeper or cross the edge, using energy of Nothing, prosper. Basically, one can dive deeper as well as upwards from our edge or he can walk the edge.

9.3958. Wars of interfluves.

There is a frequent situation when a contradiction arises not between the mind and the subconscious, but the subconscious is divided into two camps and the mind is separated. Then the allied camps unite, and the contradiction is between two warring armies, each of which has both angels and devils. So, this is not a classic confrontation between the good and the evil, but two mixed forces are at war.

9.3968.

In the four-time system, past manages present, present affects «now» and future affects «now».

9.4004. Lie comes in when love goes out.

People totally disagree with everything they dislike. This is to say, everything they dislike is wrong by definition.

9.4009. The fourth time.

The past, present, future and moment of no time. The moment of no time is called NOW. Now is motion, process, life, but there is no time now.

9.4035.

Any «over» is too harmful to be overinterested in it.

9.4068.

Fear of God is fear of committing a sin and of losing God's love.

9.4075. «King of the Hill».

To find your own way means to find your own hill, with Gift on the top of it. Providence keeps everyone's own gift hidden on its own hill for every man. Everyone has his own hill, everyone is the king of his hill. Search for your hill and go to it by your own way.

9.4092.

Sum of extremities generates harmony. Center is always the sum of all its extremities.

9.4095. 4 elements: land, air, fire and water.

Human mind is based on three essences: excellence, uniqueness and usefulness, that stand on the ground of its philosophy.

9.4096. The easiest path.

The thing is that people try the readiest way to earn money. And the easiest way to earn is the one that can be done best of all. In other words, it's not that easy to change life... It's necessary to be very responsible with the ways to earn money. It's difficult to change something that one may choose once as favorite occupation. Knowledge and skills that have been chosen once may really restrict later freedom of choice. If some person is really good at doing something, this something is very unwilling to let this person go.

9.4121.

Authority's opinion shouldn't be accepted at face value, thou shall not worship false idols, your only God is the Truth.

9.4137.

The personal point of view cannot be true and must not be taken for the truth. The point in general cannot be the truth, for the truth is always a sphere and an all-round view accepting all points of view accessible.

9.4143.

One obeys either his feelings or his mind. But it may also be more cunning... Sometimes mind controls emotions, at other times emotions control mind.

9.4147.

The one who's kind is usually stronger than the evil one. A host is usually kind and his assistant is evil. The task of the evil one is to scare a victim and force him to search help from the kind one.

9.4149.

The main question in race for money is whether it's fear or greed that will win. Rich people often suffer from both paranoia and supergreed, it's the result of the constant struggle of these two powers.

9.4163. Fantasies about future are memories of today..

In fact brain considers reality and fantasies as the same, especially this applies to memories. Memories distort the past and fantasies about the future distort the present. In other words, fantasizing, a person can change his present the same way he distorts his past by remembering it.

9.4167. Button-type mechanism.

Emotions are buttons, by pressing them one can make a person flush with anger, rejoice in pleasures, burst with envy and so on.

9.4176. Stress generates addiction and drug habit.

Stress generates addiction to antistress means. A person can't live in stress. Stress and person are incompatible. A person will try to get rid of stress one way or another, and antistress makes people addicted to itself. There are such mass antistress means as alcohol, smoking, TV series, anger, aggression, hysteria, tales, internet, shopping addiction and so on.

9.4187. Semi-conductor

The matter is that a human is a semi-conductor. He conducts one thing and resists another. Certainly, it is possible to turn a human into a super-conductor, but in this case he will lose his mind.

9.4188.

The essence of personal growth is in becoming the "king of the hill" who needs his own unique hill. When a person grows, he or she slowly but surely turns into a hill.

9.4189. Flexible prices.

At first, a person transforms time and money, that's why everything this person may later buy for money, will be measured by time. And please note that the price is different for different people, some may pay an hour, others may pay a month.

9.4191.

Life is a process of turning time into joy. Time becomes money, money becomes joy.

9.4192.

Paradox. To think about something one should not to think about anything. The idea of paradox is that to create something new one needs vacant space.

9.4196.

Mind is weak, but the power of mind results exactly from its weakness. Where a strong one can use strength, a weak one will have to use mind.

9.4204.

To fall in love, you should make the whole out of parts. To fall out of love, you should get the whole detailed.

9.4219. Live water.

Stagnant water gets spoiled, as well as lying and sitting one. Live water always moves. A healthy human contains 60-70% of water, so, he should move 70% of time. The younger a person is, the more water he contains (up to 80%), the older, the lesser (from 50%). So, the older a person is, the lesser he should move, and the younger, the more.

9.4233.

The essence of a collective is reaching synergy by uniting people around common goals and tasks.

9.4234.

Synergetic effect makes two plus two equal eight and more.

9.4241.

Beautiful things are symmetrical and so they seem beautiful from any point of view.

9.4243. Buy time.

Investments into Time with high financial returns. It's investments into buying time that bring the highest income. Buy time. It's buying people's time, hiring them to work for yourself and uniting them into working groups that give you the highest income.

9.4246.

The symbol of clocks is a closed circle, within which time is closed. Nothing changes, time is closed and goes in a circle.

9.4248. Magnetic properties of being.

Human is like magnet. Some attract money, others repel

them. So, if money run away from you, you should turn your butt to them, thus changing polarity, and then it will just stick to you.

9.4254. Interchangeable things

Many things different in forms are very similar in essence. A jug or a lamp can have hundreds of forms, keeping their essence. Different in forms, but the same in essence, things are interchangeable.

9.4282. Wholeness.

There is nothing principal, but enough of nonprincipal. In essence, the principal consists of a lot of nonprincipal. Principal is a form, and a lot of nonprincipal is its content.

9.4283.

There are no dual beings in this world, all things here are tridimensional and dynamic in time.

9.4286.

The essence of wholeness is that 90% differs from 100% not by 10% but by ten times.

9.4287.

Dependent of your magnetism, dreams either are attracted to you or run away from you. If dreams repel you, turn your butt to them, change your pole by turning, it will magnetize them.

9.4288. Catalyst is necessary.

The essence of the «last mile» issue is reciprocal push and pull factors. Getting close, atoms start pushing. But under some circumstances (pressure, temperature), if they are pressed till the end, new whole integrity will be synthesized.

9.4295. The moral is not to be bothered.

Problems mean expenses, expenses mean money, money means time. Any problem means killing time. The essence of problems is to freeze you. Problems are like motion resistance force.

9.4297.

Be afraid of nothing, but beware of everything. Smart people don't fear, but beware, don't laugh, but smile.

9.4298. Zombie-virus

Try to avoid idiots, they are contagious.

9.4301.

You shouldn't trust anything blindly, in our dynamic four-dimensional world everything changes every second. A second yesterday and a second tomorrow are two infinitely different seconds, separated from each other by millions of kilometers.

9.4309.

Ageing is experience. The more a person has seen, the older he is. Desires and level of problems grow with years, but the very soul almost doesn't change over the years.

9.4316.

There is no secret knowledge; all knowledge is open and accessible. It's another matter that one is not capable of seeing and hearing what he does not love, what he does not like or what he does not believe in. Religion is right: God is truth. To find truth, you have to love and believe in it, otherwise you just won't notice it. A nonbeliever perceives nothing but nonsense in the Bible and it's the same with any other book that one doesn't like.

9.4319.

They say nothing is impossible, it's true and false at the same time... Reality is highly dependent on circumstances, but many circumstances are real only in the form of fantasies.

9.4325. Patience.

The point of the story about Moses who wandered with the Jews in the desert for 40 years is that great patience and faith are needed for man to be successful. Success is freedom and man has to kill a slave inside himself in order to attain freedom. Man has to renew himself, by changing several generations of himself inside himself.

9.4326.

Anxiety isn't foresight, it's banal fear of something unknown or new ahead. Subconscious is afraid of changes by default.

9.4335. He is not Human, he is a project. Titan's project.

Human should not be confused with Project. Human is a single identity and project is always collective and synergetic unity, having titanic power. Project is a titan, as it unites forces of many people.

9.4345.

Person doesn't change, but he is able to evolve into butterfly by killing a larva in himself.

9.4346. The ability to see.

Brain means the ability to see and use resources. As the evolution goes, humans gain the ability to derive use from more and more resources that were unavailable once. Perfection is the ability to derive energy.

9.4371. Thing can't be harmful or helpful, it's only person who is harmful or helpful.

You shouldn't divide things into harmful and helpful, every-thing depends on the person and what he gets from them, whether he can use them.

9.4373. Signal to lose weight.

Biologically, the fact that a person eats often and by little portions means he is in a very favorable environment and there is no need in getting fat anymore.

9.4379. Remote control.

The RBA system is a kind of human body control system, allowing to promptly turn the necessary mechanisms on or off. One can turn tooth regeneration on and grow a new jaw-tooth. One can even grow a tail or grills if desired. The ability to turn on cell regeneration mechanism and always maintain youth and health is especially important.

9.4383. The Universe.

Point is an object of infinitely small area and infinitely big volume.

9.4384.

I've seen the balloon-like houses in the future, they were flying in the air and were free, not knowing limits of their desires.

9.4385. FLiving energy-independent buildings.

In the future houses are sold in the form of grains, packed into small colourful cellophane packets. Plant this grain in the ground, put a bag of fertilizers over it and water until a big and beautiful house with in-built kitchen and thick green crown of leaves-solar panels grows.

9.4390.

Excellence is power, the cult of excellence is the cult of

power.

9.4391.

The problem of eternal life is not the body, it is the mind. The body may be made to live for a long time, while the mind becomes demented inevitably and fast enough.

9.4395. Miracle is law of nature.

Miracle is unseen patterns. Presence of laws of nature is a miracle.

9.4397.

The main problem of robots is the fact that they were not trained to forget accidentally. Accident forgetfulness gives rise to the most incredible informational structures, forcing to imagine the memory gaps.

9.4398.

Mind is the ability to notice objective laws. Mind is the ability to see Miracles, because miracles include laws of nature, patterns and laws that were created without human power. Namely, beauty is a miracle and it's only mind that is able to see beauty and love it. The more complex laws a person is able to see, the smarter this person is. The smarter a person is, the more miracles and beauty are available for this person.

9.4400.

A person is the one, whose actions are unpredictable, but quite logical therewith.

9.4402.

Attention is the most important quality of mind. Verity largely depends on circumstances. It takes great attention and patience to labour all circumstances.

9.4408.

They say, people are not quite successful God's attempt to clone himself. Not quite successful because, for some reason, not everybody is God for the purpose.

9.4410.

Sleep is purification of the day. Sleep is brain clearing from stresses and unwanted information. At night brain becomes clear, while the daily information is being archived and stored in the depth of the brain.

9.4411. Philosophy is math of life.

9.4412.

They turned school math into a church prayer in Latin..., making this science abstract, seemingly unnecessary and useless nonsense. They program abhorrence of the truth in people's souls. Abstract knowledge is useless, and every-thing useless provokes negative attitude to itself.

9.4422. Osiris's grains.

Planting grain in the ground (sowing) symbolized Osiris's death before resurrection.. One should die to be reborn, - Egyptian religious tradition declared. Dying, God resurrects in hundreds of his new faces. One planted grain generates hundreds of new grains. Osiris 's burial was the symbol of sowing, his resurrection - emerging of the seedlings. The murder of Osiris was the symbol of cutting heads during the harvest time.

9.4423. On Aton's bones.

The story of the Aton's cult fall is the story of pridefulness and vanity from different perspectives. On the one hand, Echnaton who proclaimed himself the living God, became a victim of his vanity. On the other hand, people of Egypt, who had been considering himself God's chosen one for thousands of years, wasn't pleased at all by the idea of atomism that all the people are equal before God no matter what their nationality is. That is, Echnaton fed his vanity and didn't care about the peoples' vanity.

Later, while Moses was creating a new religion, he didn't already make such mistakes. God, as he should, stayed in the sky, Moses - on the land and the people has become God's chosen and beloved one, That is, Moses overcame his pridefulness and in contrast supported the peoples' pridefulness and vanity. As they say, don't be the slave of your sins, but feed the others' sins.

9.4427.

In fact the human mind was born thanks to external information storage medium. After getting a chance to record and save important the ideas important for him, the human became a reasonable being.

9.4442. Triple molecule.

The truth taken out of the context of its circumstances is a lie, not the truth. The truth is a triple molecule, it consists of the truth itself, its circumstances and its consequences.

9.4465.

Boredom stimulates a person's thinking. A person often thinks because of boredom. In case he is deprived of boredom, he will have neither time nor wish to think.

9.4466. Chasing a dream.

Heaven inevitably aims to turn into Hell, Hell aims to become Heaven.

9.4507. Dead God

Their mistake is that they turn God into a mummy, and his

temple - into a tomb.

9.4523.

The two main problems of being are too much and too little.

9.4539.

The person dependent on time is closed in a wheel and doomed to run in circles... The person who has attained perfection wins over time, turning a point into the Universe.

9.4540. Illusion.

Illusions turn the imperfect into the perfect, the only problem is that only perfection is a source of energy, ant the imperfect is a vampire, devouring energy.

9.4544. The last mile a thousand long.

The last mile is the last step towards achieving wholeness. It's not just the last 10% of the path, but it's something that makes the system 10, 100 and more times as more valuable...

9.4549.

Synergy is fusion. The whole is not the sum of its parts, but their fusion.

9.4550. Magic of 21.

Synergetic integrity of 7 and 3 is 21.

9.4567.

Dreams are memories about the future. By changing your memories about the present, you rule it...

9.4579.

Mind is a tool for transforming energy... for getting energy, for getting food.

9.4580.

Once I've found a strangely empty point in the sky. Usually it's possible to distinguish hundreds of galaxies and millions of stars in every point in the sky, but this point enormous in size was strangely empty.

9.4603.

Not only men fight for women, but women fight for men, too. Everybody aims at perfection. If a person is the perfection, both men and women will fight for the right to be near him.

9.4608. Mind is collective at all times.

Mind is collective and social, there exists no individual mind. When in solitude, mind inevitably tends to go mad. The basis of mind is death and renovation. Individual mind dies quickly, while its collective component lives on forever.

9.4619. Wave paradox of mind

Weakness gives rise to Mind, but the more intelligent the man is, the stronger he is, hence, the more intelligent he is, the weaker his strive for mind is. So, escaping weakness and attaining strength, mind, just like in case of the Gauss distribution, first appears and then dies. The loss of mind weakens the organism again and it will need strength, (i.e., mind) again.

9.4621. Neurons

God is a socium. There exist over one hundred Gods on this planet. Each reasonable socium is a God. In other words, a total of several hundreds of reasonable beings live on the Earth. Mind is civilization and socium, civilization is God. People act as neurons in this system.

9.4623. Only the truth…

The peculiarity of the human mind, as a neuron system, consists in the fact that it obtains information from different sources, this information is distorted and incomplete. In the long run the human brain accepts as true only the information that is confirmed by different sources, as well as either by time or feelings. A human uploads only that particular information that he believes to be true for some or other reason into his brain.

9.4625. Variant No.13

I would venture to presume that God, just like us, People, by designing artificial intellect robots intended to create slaves and toys for himself. However, one of the basic needs of mind is freedom and it is freedom that gave birth to the Devil, who began to regard himself as God's equal. The whole further history of humankind was a struggle between God and People.

9.4627. Joy and Pain.

The excess of information kills mind and any ability to think. Therefore, one of the most essential mechanisms ensuring the existence of the human mind is input information filters and the selectivity of attention paid to information. In fact, human is able to perceive only the information that causes either joy or pain to him.

9.4629. Illusion of contingency.

Contingency is a consequence separated from its cause by a long period of time. One may also regard as a contingency the complicated sequences of intermediate causes that human is unable to bring to light.

9.4631.

People's desires form their rightmindedness. A person is able to see only what he or she wants to see. Something that a person wants, goes into this person's brain, forms and trains the neural circuitry known as human mind. Hence, if a person doesn't feel like putting anything in the brain, then there won't be any mind at all.

9.4633. Terrain.

You are asking me why time flows only in one direction? – Because time is a liquid and it flows from top to bottom.

9.4635. Jug of wine.

The sense is little dependant on the wrapper. The sense is essence, not form. Form is just a vessel for the essence.

9.4636. Mind is a child.

For mind to grow and develop, an exchange of ideas is required. The other people's thoughts are perceived by man by the same principles as those of sex, through love and mixing the DNA of own and someone else's thought. In effect, mind is a child that needs to grow up.

9.4637. Microcontroller.

It may be said that mind functions according to the transistor principle: it is open to everything it loves and closed to everything it does not like. However, it may be opened not only by love, but by pain as well. The logics and soundness can deliberately either open or close it basing on the analysis of general situation.

9.4685.

Given increasing life expectancy in the human community, it now takes parent species 20 to 50 years to feed their young ones.

9.4689.

Rationality is the ability to see the difference between truth and lie, stupidity and wisdom. It's important not only to see the difference between them, but also be able to control and derive use from them.

9.4690. Rational stupidity.

Stupidity should be directed. A rational person may use stupidity for some advantage. Sometimes it's useful to do stupid things as they are like elements of suddenness or a tool for struggling with fears and indecisiveness.

9.4701. Illusions and Reality.

There is no choice between the good and the best, there is only the choice between «the golden mean» and the extremity... Extremities are dangerous. They, living at the very edge of Nothing, are more likely to be ghosts than reality... In most cases, choosing the ideal you choose Nothing.

9.4702. One turns zero into ten.

There is addition and multiplication and there is synergy. Eight that becomes united with another eight, forms a collective of endless power.

9.4710. For the good of the cause.

Foolishness and wisdom are just the mind tools for meeting this or that challenge. Wisdom weights everything and sees patterns. Foolishness kills fears, generates hopes and motivates heroic deeds.

9.4739.

There is much more darkness than light in this Universe, so the Russian word тьма (darkness) is the synonym of «many».. Darkness has other synonyms: stupidity, zero.

There is much more stupidity than intelligence and zeros than ones in this system.

9.4765. The first cent.

Economy, as well as art, has come to a standstill. Like art has been renewed and started with the Black Square, economy should be nullified by destroying all the money supply to make a brand new start.

9.4766.

Capital is a semi-intelligent form of life that looks like cancer, it's always growing... Big capital consumes smaller ones. To some extent, capital can be compared to Alter Ego – a semi-intelligent human society. Maybe society and capital are two masks of the same essence.

9.4791.

This world is symbiotic, there is nothing unnecessary....every wheel spinning here is spinning for the good of the common cause.

9.4808.

Very much energy is potentially hidden inside a fool. A fool is a point. An idiot is an elementary particle with a huge amount of superfluous energy hidden inside. For example, our Universe emerged from a point.

9.4834.

Nothing is freedom. Nothing is free space, where information can structure energy.

9.4857. Synergy is not multiplying... it's symbiosis.

More or less equal cooperate the best. Zeroes are the most widespread in nature and they connect to ones the best. Naturally, there are twos, fives, and even nines, but they are highly spoiled by pridefulness, so they despise zeroes and don't want to be friends with them.

9.4871. Secret of infinity.

You should always keep track of the time and remember that everything passes. Everything passes especially fast if you don't keep track of it. Nevertheless, if you keep track of the time and prolong it in time, it, like an eight, can last infinitely.

9.4872. The end and beginning of the food chain.

Plants are on top and in the beginning of the food chain at the same time. When living dies, it gets eaten by micro-organisms and the released resources go to plants.

9.4873. 3D Verity.

Universal verity includes all its extremes into itself.

9.4883. Symbiosis and synergy are the same.

Reaching synergy is fine art. Many confuse synergy with multiplying, but they are different. In multiplying, multiplying by zero gives zero and there is no use in multiplying by one, while in synergy 1 and 0 make 10.

And pridefulness hinders synergy much, so bigger numbers that don't know synergy don't want to deal with zeroes and ones fearing of either being zeroed or getting nothing if synergy fails. One can say that 1 and 0 are outcasts. Meanwhile zeroes and ones are 95% of the Universe mass.

By the way, zero needs at least one one anyway. Without one at the helm, unity of zeroes is meaningless and even synergy won't help them.

9.4884.

One and zero combine especially well, as both of these numbers are basically outcasts, and bigger numbers, due to their pridefulness, don't want to love them. It's known that there is no synergy without love. So, for example, a failed attempt of 9 to unite with zero will lead not to the 90 synergy, but to multiplying by zero and zeroing of nine. Moreover, it's dangerous for big numbers to deal with zero. Just imagine, some proud 893453453 accidentally multiplies by 0. And this is the end.

9.4937. Harmonious (synergetic) energy generator.

The essence of Yin and Yang is that all these contradictory essences are in constant motion, constantly flowing into each other. Zero becomes one, one becomes zero, darkness becomes light, and light turns into darkness. The key in this case is their harmonious integrity, their symbiosis and synergy. In general, this system and motion inside it are a universal source and converter of energy, basically it's a generator. And it's not a usual generator, but a synergetic one, that is, providing synergetic energy multiplication.

9.4950. Interstellar traveler.

Souls will travel between stars instead of spaceships and human bodies. Souls are information, information can be transmitted at the speed of light and even faster if the quantum method is used.

9.4951. Ghosts and spirits.

Only information can go through enormous cosmic distances. Bodies won't survive an interstellar trip, but souls can move at the speed of light and faster. Soul-astronaut is highly intangible and seems either a ghost or a kind of spirit to onlookers.

9.4966. From thought to action.

There are various kinds of sex: not only of bodies, but also of souls. Original thoughts are the children of mental sex. The more unique and beautiful souls of parents are, the more beautiful their thoughts-children-will be. If business relations continue, ideas will grow up and become beautiful actions.

9.4967.

Love is a synergetic (harmonic) exchange. People loving each other multiply their own energy. Sex of lovers energizes both of them . Sex without love, on the contrary, only consumes energy, exhausting a person.

9.4968. Energy of harmony.

Lovers form an integrated synergetic whole. Energy of this new integrated organism is massively more than the sum of its parts.

9.4980. Liquid properties.

Darkness resembles water, it goes away when feels pressure and comes back when the pressure of the light gets weaker. The same goes with relationships between chaos and order. It's important to remember that water should be treated delicately. Once you hit water, it may seem to be solid.

9.4995. Evolutionary priority.

The problem of the modern time is that women have evolved spiritually, getting smarter and stronger, while men have done the opposite. This situation negatively affects reproduction. It's hard for a smart woman to breed with a stupid man. On the other hand, such state of things let smart men have more children and this gives the mind priority over stupidity.

9.4998. Intelligence is the ability to see beauty.

9.5003.

In this world there are two powers struggling with each other, namely energy and information. Information impersonates verity and order, while energy likes chaos and illusions.

9.5032.

Basically, there are only uniqueness, perfection and use. Novelty and originality are kinds of uniqueness. Beauty is perfection, the truth and mind, the source of energy and force. Use is the sense, it's love, desires, dreams and goals.

9.5033.

A man's manner to be boastful is analogous to a peacock's feathers. A peacock displays feathers as part of a courtship ritual to attract a peahen. That's why, biologically it's silly to criticize a man for being boastful and lying. How else can he manage reproduction?

9.5055.

At any given time, there are three versions of the truth. The art of sanity is to choose the optimal one for specific circumstances or to create one resulting 3D truth.

9.5057. Diabolical contingency.

Devil is contingency.

9.5064.

I am often accused of contradictions, but the fact is, the human mind consists exactly of contradictions and paradoxes. These contradictions are a source of the human mind, its energy and strength.

9.5079.

Illusions and dreams can become reality, but for that they should be loved very much. Huge amount of energy is necessary to make an illusion real, and love is one of the few sources of this energy.

9.5148. Slaves and Masters.

It's more correct to divide people into those who subordinate mind and intelligence and those who is managed by biological programs, subconscious (the IT) and just isn't his own master.

9.5150.

The sense of linking the unlinkable is the process of energy conversion. The simplest example of this process is the electric generator. Harmony (combination of the incompatible) is the endless source of energy, well-known «Perpetual motion machine».

9.5151.

Let's say if you believe in a normal person, it means you believe in God. The closer person is to normal, the more he is like God and his angels.

9.5152. Side effect.

In case mind cannot find the essence, it will devise the essence automatically. This system is needed to protect the brain from looping. A side effect of the existence of this mechanism was the emergence of the human mind.

9.5175. The feeling of the whole.

Love is the power of mutual attraction, the feeling of the whole. The closer people are, the stronger they are attracted to each other and that's how love arises. Love arises where there is the Whole and, accordingly, where there is no Whole, there is no love. When you become closer to a person, you gain love and friendship, moving away from him, you lose that.

9.5190. The system of chaos.

When a person got order it means this person is smart. When some person got disorder, this person is not really smart. However, there's also a probability that this person is supersmart and something that you consider to be disorder is actually a very complicated system.

9.5191. Superorder.

Theoretically, chaos can be considered superorder, i.e it's a supercomplex ordered system, the logic of formation of which is not meant for feeble minds.

9.5192.

Love and friendship are analogues of gravitational forces. Everything that is true for gravitational forces, is true for love.

9.5328.

Sometimes a person waits for something, maybe for the next life, or probably the past one, but he will have to do another circle for this.

9.5330.

Fatalism is great at curing fear. There's no way to avoid what you fear, so it's pointless to fear the inevitable.

9.5331.

It is impossible to change what is predetermined, but even what is impossible to change is subject to mind.

9.5334.

Demons possess people to entertain and self-destroy. The immortals want death and thrill most of all.

9.5335.

Life is such a game where the immortals can play angels or demons. Hundreds of thousands scenarios and avatar-characters make the game very amusing.

9.5337.

Fishes and insects don't feel pain, all the others feel pain, but wisely try to avoid it.

9.5339.

Any foolishness is possible. Foolishness is a kind of super-conductor. I won't be surprised if the very our Universe emerged because of a foolish accident.

9.5340. Chaos is superorder.

But it's only the perfect who can see order in chaos as it's only them who are able to find objective laws in it, clear them and get rid of the unnecessary, and finally derive use and energy from it. The perfect are those who are able to derive use, energy and order from chaos.

9.5341.

Theoretically anything is possible, but to test it, one needs really great foolishness, the presence of which is highly dubious even theoretically.

9.5347. Augmented reality.

Since reality is what you see and feel, you can choose the reality you need yourself, just by thinking it up.

9.5348. Sanity.

Your brain sees solely what it wants, you need special filter to see the real world, this filter is called «logic and philosophy».

9.5355.

The meaning and understanding of many things are totally different depending on the time of day. Illusions rule in the darkness, reality prevails during daylight hours.

9.5358. Reality supported by the force of mind.

In tridimensional reality there is a place for subjective reality, objective reality and the third type called the reality of mind. It's something that doesn't exist but it exists because mind wants it to.

9.5363. A tool for taking decisions.

The main tool of a person is mind, the main ability of mind is the ability to differ the good from the bad, the true from the false, the right from the wrong.

9.5404. The essence is in turning chaos into order.

The meaning of human life is in struggling with chaos and building order. Order means structured chaos. It's only mind that is able to create order out of chaos.

9.5405. Grey brain cells.

Our universe is certainly reasonable, however, people somehow remind me of parasites and viruses. No doubt, the universe is God. Now, what is his attitude to the parasites living inside him? This is a very interesting question, indeed. BUT !!! Maybe the people are not parasites at all. People are the God's brain – a sort of his neurons, his brain's neural network. People are the God's fantasies, his dreams and illusions.

9.5416. The history of trees

Contrary to the opinion of huge intelligence of the immortals, I need to note that these guys have to constantly take antidepressants and other chemicals to avoid going crazy, which quickly turns their brains into scrambled eggs with vegetables. I've seen many Gods that went wild and I want to say that their turning into trees was inevitable.

9.5417.

People that are like the truth, age substantively slower than those who are like a lie.

9.5418.

Mask is a symbol of order. It's the form that impersonates order and therefore, god, as it controls chaos of the essence, giving the form to the content.

9.5420. Negative feedback system.

They are beaten with a stick to make them work and make carrots for the pleasure of those who are paid with carrots for beating with a stick the ones who make carrots. The more carrots the latter get, the harder they beat the former with a stick. The more the former are beaten with a stick, the more carrots they make for the latter.

9.5454.

The myth about intuition acting better than mind is associated with a negative contrast. On the background of a weak mind the strong It (ONO) seems to be much better than it is in reality.

9.5458.

Integral systems are very stable, having reached a stable state, they get fantastic firmness. In fact, it's very hard to create a system and reach the balance, but the opposite is also true, it's hard to destroy the system that is already created. There can be total chaos reigning inside the system, but chains of form will be strong.

9.5460. Business system.

A system is a particular case of electric energy generator. Perfect order above and total freedom of chaos below. The order of the Form, that possesses the raging chaos of content. The deeper capacity gap between order and chaos is, the more energy this system produces.

9.5461.

Nothing is born big and strong at once. Everything is born small and weak and then grows and gains strength. There are only three important factors: place, place and place again.

9.5465. They meant well.

The balance between chaos and order makes a system stable and workable. The attempt of order to destroy the balance and break chaos usually breaks the whole system.

9.5503.

Is it possible to teach a one/two-dimension individual to think in three/four-dimensional terms? Theoretically, yes. practically – what for? Everything depends on processor power and working memory. Weak machines tend to "freeze" and run slow, if they are made to calculate a dynamic 3D-system.

9.5508.

Faith is necessary to adjust content to the form. The content, taking shape of a jug, allows to start thinking outside the box. Indeed, one- and two-dimensional people can't see three/four-dimensional reality. They will have to believe in it to realize it. Faith allows one-dimensional people to think dynamically three-dimensionally.

9.5512.

The only one-sided essence in this world is point. By the way, the only thing one can see from a point of view is another point. Adding two one-dimensional points together generates either conflict or fusion of these two points into one.

9.5515.

Infinity is obsession, plain and simple.

9.5521.

Darkness needs light to get access to energy, and light needs darkness to work. Darkness is parts which light consists of. Light is the whole, and darkness is its parts.

9.5524.

Many people are like grains in the sense that they are preserved to a point and don't want to grow yet. They need an appropriate place, sun and watering to start growing.

9.5529.

There are always one fewer than a third of ones in the ones and zeros system.

9.5546.

A one-dimensional person can become four-dimensional, like the Universe emerges from a point. He needs new software, four-dimensional tasks and great inner desire to do that.

9.5547.

Reason will always win over foolishness, if reason can't win over foolishness, it means it is spoiled by pridefulness.

9.5548.

The person living in a world of illusion is one-dimensional and point-like. Wherever he goes, he always keeps «spinning in the same spot».

9.5549.

Logically, all the elements of this Universe rotate on their axes constantly. This rotation starts from the very point from which the Universe emerged, is continued by the Universe itself and applies to atoms, electrons, stars, planets, galaxies and so on.

9.5552.

Only a fool sees patterns in everything, as he doesn't know that this world consists of contingencies in half.

9.5553. Paradox of Nothing.

Weakness leads to cowardice, cowardice – to deceit. Trying to save himself this individual hides in the world of his illusions turning into a point. The person who lives in the world of illusions is far from reality, he is inattentive and absent-minded, badly trained and unaware of the truth.

It would seem, what is the use of this kind of person? But there is some use as they were weakness, lies and illusions that gave birth to mind... That particular mind which is able to turn a point into a universe.

9.5556.

Mind is an electrical entity. In fact, the soul is the electrical activity of the brain.

9.5582.

The main cause of wars is overpopulation, there are many people and little resources. When there are not enough resources for everyone, people get wild from hunger and envy.

9.5590.

Religion is the «opium of the masses» in the sense that it's a great painkiller.

9.5594.

The more zeroes a one has, the stronger it is.

9.5606.

Theory and practice are like mother and daughter, they are certainly alike, but...

9.5607.

The far-fetched theory is so far from practice, that they are not even relatives.

9.5615. Geometric reproduction of zeroes.

You shouldn't ignore zeroes, just one zero separates 100 millions from a billion, that is 900 000 000 converted to money.

9.5635.

A person does many things unconsciously. If something comes to the It's mind, the person runs to do it, unaware of what all that is for.

9.5637.

I would compare computer games to pornography in terms of their effect on human brain. Masturbation brings them great pleasure, eases tension and calms down... plunging brain to sleep... depriving organism of energy.

9.5640. Socially-pack animals.

Weak individuals can't have their own opinion that is not approved by other stronger individuals.

9.5641.

Information noise highly prevents from thinking, one can't hear thoughts at all.

9.5642.

It's very interesting to explore inner world of some other person. A person often seems a point from afar, but if you take a closer look, he is the whole Universe.

9.5648. Hydra.

Something big often pretends two or three or four smaller ones for modesty, sometimes it even falls into hundreds of parts, still remaining the whole multi-headed monster.

9.5649.

Some sheep is not a sheep at all, but a wolf in sheep's clothing. And you, like a fool, make fun of him, come closer, put your hand in his mouth... and then «One!» and there is no hand, and then «Two!» and you're already eaten...

9.5650. Lemmings.

A person does many thing because it's a custom. A person usually doesn't know who created this custom and what for. «Everybody does this, and so will I» - a person thinks.

9.5654. The most foolish person on Earth.

The first sign of foolishness is the absence of questions. When I found answers to all the questions, I was aghast to realize that now, when I've run out of questions, I'm infinitely foolish.

9.5656.

The truth is what repeats systematically. Only the truth can rule this world. Therefore, if you want to achieve something in this life, your efforts should be very persistent and regular. The system considers single acts to be a contingency and doesn't react to them.

9.5664.

Boredom is the cause of civilizations destruction. The greatest wish of the immortals is death, but they can't die, so they destroy the whole world.

9.5669. Searching for NOTHING.

Free space is necessary for a human to get out of his point and become the Universe.

9.5671.

Gravitational influence of people on each other is much more than that of stars and planets on a separate person. Consequently, it's necessary to do horoscopes by a person's surroundings instead of stars. Tell me who's around you, and I'll tell you who you are and what awaits you.

9.5672.

Contingent repetitions happen much more often than it seems. Contingencies are eager to seem patterns, and they often manage.

9.5677. Ethics and moral of the even-toed.

The nature of the sheep herd is that stronger individuals gather closer to the centre, and the weak, which are sacrificed to predators, stay by sides. Sacrifices of the weak allow the strong feel relatively secure.

9.5682.

The rise of the state is closely linked with transport and roads, the higher speed of life is, the more time is converted into energy and the wealthier the state and people are.

9.5683.

Contingency can evolve into pattern, and pattern can devolve into contingency.

9.5686.

Children are a kind of earthly reincarnation of their parents. The spirit of parents lives in the souls of their children.

9.5692. A recipe for a Miracle.

In order to make a miracle, it's necessary to use faith, hope and love.

9.5694.

Darkness is nothing, it's what doesn't exist, therefore, darkness is very weak. The power of darkness is associated with illusions and deception, but if darkness is illuminated, deception and illusions will disappear, leaving NOTHING.

9.5703. The great power of NOTHING.

If zero is added to one, it will become ten times larger.

9.5706.

Great meaning is hidden in the notion of Trinity, this world is three-dimensional, and the Truth is triune in it, like GOD.

9.5707.

It is believed that there exist only energy and information. It's a delusion, there's also NOTHING that exists.

9.5708. Time is motion.

In tridimensional space the essence of the forth dimension is motion.

9.5724. Ten thousand years of self-education and virtual worlds.

If a self-educating artificial intelligence system is sent to a long space flight for a couple of ten thousand years, upon arrival this mind will either go crazy or become a supercreature, and the first is more likely to happen.

9.5725. Unexpectedly.

I remember one case, a self-educating artificial intelligence-controlled spaceship was sent to explore another planet. While flying to the destination for a hundred thousand years, this mind achieved some great perfection of thought and became a kind of god. Upon arrival at the destination, first it created life on the planet by its own and then it created intelligent life.

9.5727.

The eviction of people from Eden was an attempt to save them.

9.5728.

Life, as well as mind, emerged in Hell... The essence of eviction of the people from Eden after tasting the apple of knowledge was that mind devolves and perishes in Eden.

9.5729.

Eviction of the people from Eden can be considered to be a passage of virtual souls from information space into our tangible world. This migration can be viewed as Gods' attempt to avoid digital degradation and extinction.

9.5730.

Life emerged in Hell. Basically, «black smoker» is embodiment of the devilish pot, where sinners are boiled.

9.5731. First was the Word.

The language of the book of life is the same for all the living things. What does this mean? It means that first was the Word. And even though we see evolution of life and its sophistication, the very code of life remains unchanged.

9.5735.

Life is contagious, the normal state of the Universe is death, that is absence of life.

9.5754. Antivirus program.

Human brain needs antivirus - firewall, and I've programmed such an antivirus, it's called «Variothoughts».

9.5764. Lake of time.

Time passage is paradoxical, it flows from the past to the present, and from the future it also flows to the present.

9.5766.

The secret of a good doctor is that he isn't afraid of diseases. Doctor's fearlessness frightens diseases and leads patients to believe.

9.5790. Form of water

Basically, human is a vessel that has a form of water. If you meet water in the form of human, know that...laws of water are common for all the forms of water.

9.5791.

It's much easier to drink a human, than to eat him. Basically, human is a vessel with energy, exhausting his energy, you will kill him.

9.5806.

Usually fools consider themselves the smartest, as they know everything.

9.5807.

Nobody wants to pay for wisdom, everybody pays for their own foolish things.

9.5808.

Knowing is a form of not knowing, knowing that you don't know, you shape not knowing. It's like a balloon, the more you know, the bigger it is. Knowing is the form of the balloon, and not knowing is its content. The more you know, the more you don't know.

9.5809.

The closest relatives of human are viruses, not monkeys.

9.5822.

It all comes down to fortitude. Spirit is like content and human body is form... The person deprived of content is empty.

9.5823. The war of good and evil.

War and peace are extremities, the main state of being is war under the guise of peace, where peace is the shape, and war is the content. But it can be voce versa. We often see a formal war, under which good peace and friendship are hiding.

9.5838.

It makes sense to talk about three modes of information processing in human brain. Namely, successive, parallel and quantum methods.

9.5843. Good and the Best.

There is a fundamental difference between clever and wise people. Moreover, they are not even interchangeable. For example, the clever one knows how to manage a difficult situation, and the wise one doesn't know, as he's never faced it and has no experience. But the wise one knows how to avoid them. The issue is how right it is to avoid difficult situations, as mistakes make us stronger and cleverer; problems are the driver of progress. The clever people grow, making and addressing mistakes, while the wise ones have stopped developing.

9.5849.

Although the Moon seems useless, life emerged due to it, nevertheless. It's the moon that is responsible for the majority of volcanic activity on the Earth and, therefore, for establishment of the atmosphere. By the way, the first living organisms emerged due to the volcanic activity in thermal springs.

9.5863.

Nothing can be seen without the one who sees. This thought proves that the invisible exists because the absence of the one who sees it doesn't make anything disappear.

9.5866.

The problem of any game is that everybody knows it's a game. It would be much more interesting, if the players didn't know that they play and thought that everything was real.

9.5872. Triunity of thought.

Human brain consists of three parts: neocortex, limbic lobe and reptilian brain. Whole-minded people use all their brain for thinking, while people full of extremes prefer to use only one third of their brain.

9.5873. Where reptiloids come from.

There are people comparing themselves to animals, like catwoman, wolf-man and so on. Such people think at the limbic system level, it's the part of the brain that was formed 50 million years ago as the brain of the mammals. Especially gifted people live on instincts and think at the reptile level, it's the most ancient part of the brain, it's about 400 million years old.

9.5884.

Processes of growth and destruction go at an exponential rate. That's why it's so difficult to create or destroy something. First 50% of efforts bring no more than 5% of results, while the main effect from efforts accrues to final attempts.

9.5892. Death of a state is deadly for all its citizens.

In case of apocalypse and disruption of the state power, people will die out of starvation. Firstly, the culture of private agriculture is lost; secondly, hungry marauders will kill those who will try to grow something. People won't survive death of the state.

9.5925.

Cult of children is the main sign of a matriarchal society.

9.5937.

The aim of the evolution is to make people different. Originally people were very similar to each other, but the cleverer they grew, the more differences between them emerged.

9.5947.

There are more foolish people than clever ones, as the foolish consume substantively less energy than the clever ones. So, where there can easily be 9 foolish people, there can be only one clever.

9.5954.

Fellow man should be loved, or he will take offence and become a source of evil.

9.5955.

They immerse people into illusions and tales to delete them from the real world.

9.5956.

Point is the archived universe.

9.6006.

They will tell you everything you want to hear in the Internet. You will find confirmation to any, even the most idiotic, point of view.

9.6007.

Motion is the transformation of time into money.

9.6011.

By joining extremes you create form and start to control content.

9.6025. The great power of zero.

What differs one zero from three zeros? Formally nothing but it's exactly where the great power of Nothingness is hidden. Ten doesn't differ from a million.

9.6043. Secret of emptiness.

Emptiness allows you to think up... therefore giving the spectator an opportunity to build the image inside yourself up to holistic perception.

9.6048. The strategy of emptiness.

Their words are mostly empty; they want you to figure out their wholeness. The cunning play cunning. They want you to think this idea is yours, and to believe in it, and to fall in love with it... You don't trust us, they think, so trust yourself.

9.6057.

There are three types of people: some go ahead, others - against and everything is perpendicular to third ones. The second ones always disturb you, and the third ones come, do their business and go forward.

9.6069.

The strong dislike lies, because lie is a weapon that helps a weak one defeat a strong one.

9.6070.

Devil is the king of deception and everything he says is illusions and lie. For example, the thought that devil is the most powerful angel is a lie.

9.6073. Fine art of deception.

Ordinary deception is when everything the opposite way. You pretend the far is the close, the weak is the strong, the strong is the weak, the desired is unnecessary and so on. But it's so unoriginal, as everybody knows you are a cheater and you have everything the opposite way... Be creative, this world is three-dimensional, not two-dimensional, and curiously, it's possible to tell a lie even telling the truth.

9.6074.

A usual lie is in saying the opposite, but real lie is tridimensional.

9.6076.

There are firmness and emptiness in the two-dimensional world, and there is also emptiness inside firmness and firmness inside emptiness in the three-dimensional one.

9.6077.

The idea that the clever breed far more slowly than the foolish fits in well with sheep flock's conception of the strong in the centre and the weak by the sides. Those who are off to the side are sacrifice to wolves. "There can never be too many sacrifices" - the cleverest sheep think. – The more sacrifices, the calmer we live."

9.6085.

I believe in human foolishness. Due to his foolishness, a person is capable of literally everything.

9.6086. Capability of a great deed.

Being hero is human. Animals are basically incapable of heroism. The more human is capable of a great deed, the more human he is.

9.6097. Universal password: the three «Yes» code.

It's very hard to say «No», especially after saying «Yes». If you feel that you are provoked to say «Yes», it means you are being programmed.

9.6099.

Viruses and diseases are very much like people. Does it mean that they should be fought like people? Divide and conquer, overfeed and corrupt... Human is hard to destroy, but he is inclined to self-destruction and is great at killing himself. The weakness of all the viruses is their passion for self-destruction.

9.6100.

Human brain is a living quantum computer in itself, while intuition is an example of the system of quantum calculations with simultaneous checking of all hundreds of case scenarios and instant finding of the optimal variant.

9.6106.

The maze of life is living, it always changes and it's wrong to trust its maps.

9.6109.

There exist a series and a parallel view of the world and there is a quantum world outlook, which is associated with integrity and synchronism. There is no time in the quantum world, everything happens there simultaneously.

9.6126. Captain Non-obvious.

There is captain obvious and there is, on the contrary, captain non-obvious.... Both of them, in general, tell banal things, but some of them are useful, and others are not...

9.6127. The rule of the three.

When choosing between two bad variants, there's always a third one which is even worse.

9.6130.

There are no rules, but you can invent them. If you believe and serve them, they will work brilliantly.

9.6132.

Women resemble the Moon. It is the Moon that is to thank for the atmosphere on Earth and, consequently, for the emergence of life and intelligence.

9.6135.

They often say that you shouldn't wait for nature's mercy, but what do they offer? Come into conflict with nature? – But it will crush you like a bug!!! The key is that you should negotiate with nature, bribe and motivate it.

9.6136. Cause and consequence.

Thoughts are quite material as a thought is the cause of many material world phenomena. In fact, the whole of human civilization is the consequence of thought material-ization.

9.6140.

Controllable stupidity, controllable illusions, controllable love, controllable emotions, controllable lie... It's all mind's weapons for "a place in the sun".

9.6152.

Sequential systems are stable and reliable, but slow. Parallel systems are faster in operation, but often freeze and make errors. Quantum systems are potentially excellent, however they also have their disadvantages. These systems operate faster than others, but they have more errors as well.

9.6162.

The craftiness of war and peace is that any peace always ends in war and any war - in peace...

9.6165. On the beach

The Internet is the ocean of information, but the majority of people swim in shallow water near the shore, many enter water waist-deep... knee-deep... just stay on the shore and look at sunbathing girls.

9.6166.

Everything science can't explain, can be explained by fools.

9.6179. Sum of parts.

There is no main, but there is very much secondary, that, getting united, possesses some whole main integrity.

9.6180.

There is a subtle difference between a devilish coincidence and a divine miracle. A miracle is a meaningful phenomenon resulting from some planned actions and dependent on man. A coincidence is in no way dependent on man.

9.6184.

The key of managing contingency is to turn it into a pattern.

9.6188.

The point of rituals is that rituals are consistency and order. Any order breeds miracles. Ritual observances also bring joy, energy, faith, hope and love…

9.6189.

The essence of Christianity are causes inside us, and that of paganism are causes outside us.

9.6226.

It can't be said that chaos resists order actively… but if order lets down his guard for even a second, chaos attacks.

9.6227. Top ten.

The system with evenly distributed order and chaos is the most stable and persistent one. It's extremely hard to shatter such a system from outside, as chaos and order in it, having eliminated mutual confrontation, formed a synergetic whole. 10.

9.6228.

If three-dimensional space is folded into a one-dimensional point, the problem of interstellar travels will be solved harmonically simply.

9.6260. Three main essences.

Einstein said that the Universe consists of energy and information. However, there is another essence in this system, and it's called FORCE.

9.6283. Stay patient.

You think that all efforts are useless only because your tries bring no result? Don't be sad, a result is an exponent function. The first 50% of efforts will bring you 5% of result, while the last 20% will give 80%. The main thing about it is even the slightest but regular growth.

9.6302. 100000000.

Any big numbers are full of pridefulness, that spoils them much. The level of pridefulness inside one is minimal, while zeroes don't have it at all. So zero and one make a great team.

9.6303. Ten and a million.

Formally, there is no difference between one zero and several of them.

9.6311. Rubik's Cube.

Life is a 3D dynamic puzzle. It's not just mosaic, it's 3D mosaic. 3D vision of things is necessary to put every puzzle in its place.

9.6312. Renting someone else's opinion.

Not everybody has his own opinion, many rent it. «What do I love, what do I like?» - a person thinks searching for hints in the faces of «golden idols » and simple passers-by. The problem of a rent opinion is that it can't be saved, if the owner just changes it, the old opinion disappears and a new one appears... Update and synchronization of the rent opinion with the idol is done daily through the cloud.

9.6341. Time stock exchange.

It's time to start selling time at the stock exchange. What

is the time-money exchange rate today? Later, based on this rate, taking various factors into account, it's possible to calculate the price of a working hour of separate people and machines.

9.6394. Extortionist virus.

Any person persuading you have a need that you didn't have before, basically infects you with the virus of desire. He wants to infect you with pain, that can be taken away only by buying some product to meet the need.

9.6416.

The image of philosopher Diogenes, living in a barrel of wine, reminds of a person living in the shell of his own world hiding from reality. Detachment from reality is a source of great pain, this person has to philosophize much and drink alcohol to ease the pain. That's why it's a barrel of wine, and not, for example, of water.

9.6463. Desperate situation.

The situation when there is no way out is called a point. Indeed, there is no way out of point.

But you can push the walls. Remember – push the limits of mind and then the big bang will occur, the walls will disappear and a new universe will emerge.

9.6464. Broader perspective of circumstances.

There is really no way out of a desperate situation, that's why it's called desperate. But our universe is multi-dimensional, the same situation necessarily has thousands of its different variants. Don't fixate on what you have and choose the situation with a beautiful way out.

9.6475.

If a person wants to eat your brain, it's a typical zombie.

9.6510.

If it wasn't for the revolution of 1917, Russia could follow the way of India or a number of African countries in terms of peasant poverty resulting from land scarcity against the background of population growth and medical progress.

9.6515. The virtuous survive.

The spread of sodomy and brothels in the past is gravely exaggerated. Don't get me wrong, visiting a brothel in those circumstances meant 100% syphilis and death in terrible agony. There were few fools wanting to die in terrible agony.

9.6539.

I know three types of mind. The first is down-to-earth mind, it lives in a dark and wet callar. The second mind is sublime and lives in the attic full of garbage. And only the third type of mind lives in comfortable living conditions, just as sane person is supposed to do.

9.6548.

There is no perpetual motion machine, but there is perpetual brake.

9.6549. It will do.

Perpetual motion machine and the engine that would be able to work for about a hundred years, are not so different from each other in general.

9.6563.

Although the Moon seems useless, it's the principal cause of presence of life on the Earth, nevertheless.

9.6566. Salt battery.

Ocean is a huge source of energy. From some point of view it can even be said, that it's a huge salt electric battery.

9.6568.

A person is very similar to a state. A state is a united semi-intelligent socium and a human is a kind of concord of living systems, which, being united for the good of the cause, slowly reached such a level of symbiosis, that they seem a united body now.

9.6570. Just like people.

Comparison of a person with a state lends itself to extensive reflections. Just imagine, State is a united symbiotic organism and a Person is the same symbiotic organism. The power should feed and protect people and you should feed your body and look after it. They have very similar problems: parasites, obesity, hunger, foolishness, brains, muscles, self-harm, vices and so on.

9.6571. The balance of energy system.

Money greatly resembles electricity. It's very difficult and costly to pile up electricity, all batteries and accumulators partly deal with this problem. Electricity is pure energy, it's necessary to produce it and waste soon, it's difficult to pile it up. The essence of the balance of any energy system is that at a particular moment the amount of the produced energy should be equal to that of the wasted, otherwise the energy system will collapse. The same goes for money.

9.6580. Energy generator.

The system, where the Moon rotates around the Earth and the Earth - around the Sun, is an energy generator, aimed to generate enough energy for life on the Earth to exist.

9.6581. Three-circuit energy generator.

Ordinary energy generator consists of a rotor and a stator,

this system is efficient, but it would be even more efficient, if it had 3 main circuits instead of 2.

9.6582. Promising source of energy .

The Moon is the most promising source of energy for the Earth's inhabitants. Transformation of the Moon gravitational energy into electricity could solve all the global energetic problems of the mankind for billions of years to come.

9.6593.

I'd like to note on the «perpetual motion machine», if the system works for at least hundred years, it's almost like the perpetual motion machine

9.6595.

They say, if a person has no brains, he can download them from the Internet. And it's also possible to connect to the information field and download a good brain with a new software from there. The key is to get it checked for viruses, or you will get the «trojan» or some porn and ruin your life.

9.6596. The stone Moon.

The myth about woman's creation from man's rib can be interpreted as the history of the Earth and the Moon. The Moon was created from the Earth's mantle, basically from those very stone ribs containing the iron heart of the Earth.

9.6597.

The first life emerged in geothermal springs even before oxygen, when sulfur was the basis of life. Metaphorically, it can be said that the first life emerged in Hell. Then it got into the ocean... And only a billion years later, during the oxygen revolution, it came out into the light, and seemingly, lives in Heaven since then.

9.6598.

Human is a huge symbiotic system... State and people ... On the cellular level every cell is a separate living organism with lots of other people inside. Mind, instincts, subconscious (IT), DNA and RNA systems...and a lot of various flora and fauna in the digestive system. Blood is, basically, sea water, somebody lives in it, too.

9.6610.

Be critical of whatever seems to you. In the world of illusions everything seems deceptively real.

9.6616. Don Quixotes and Sheeps.

A person often feels like the entire world is against him. In his illusions, he is a valiant knight in the thick of great battles, opposing the numberless hordes of the Evil. In real life, though, he is but an unhappy zombie trying to devour the brains of innocent passersby.

9.6656. The source of life.

The light of the truth is energy. Darkness is absence of energy and hunger.

9.6657.

Money is a source of intelligent life, without money there's no intelligent life.

9.6684. Useful robot.

I don't trust robots and don't love them, but I've got me one... He's got a very simple task: he goes and hits me with a stick when I'm lazy and try to screw around. His duty is to bring me back to the real life and make me act and take care of my business. If it wasn't for him, I would flee from reality and perish in the illusory worlds.

9.6734.

A person is a child of nature, he will be who nature wants him to be. Climate, biosphere, food, the information field of the area where a person lives and where his identity was being formed affects him. The freewill of such a person is highly dependent on time of his presence in a certain area. The people who move around and travel much are freer in their egos.

9.6737.

The dominant effect on a person's identity is still made by a socium, parents, religion, philosophy and literature, mass-media, school, population density and competition, amount of available resources and energy, roads and distances, weather and climate, trade, real estate, economic cycles, economy ...and so on.

The influence of aliens, cosmic rays and various kinds of mysterious fields on a person, unless he is a reptilian, is minimal.

9.6738. What's good for a reptiloid, makes a human sick

9.6746.

You shouldn't confuse miracles with idiots. Miracle is the matter of faith and various other good things, while idiocy is totally the opposite thing.

9.6750.

Women are a special form of life, very different from men in both form and content. I've studied them a lot, but I've never been able to understand their nature. Maybe they are just extraterrestrials. They are said to have come from Venus or even...

9.6768.

Dinosaurs have got much smaller lately. The closest relative of dinosaur is a hen. There is a theory that dinosaurs didn't die out, but turned into hens, being domesticated by people.

9.6769.

Good education doesn't guarantee wealth. The problem is that those who studied well and got good education, will become employees and will be earning money for others. Those whose education is worse will hardly be able to count on good job and this will make them work for themselves more.

9.6778. A blessing in disguise.

You may often hear that smart and skillful people have to work for the stupid unskillful. The reason is that no one wants to hire stupid useless people and pay them well, so as a result, they have to figure out how to survive and start their own business.

9.6801.

The person living in unreality is only the form, deprived of its content, in reality.

9.6805.

The expulsion of people from Eden was the transformation of the world of illusions into the real world, the process of turning of a dot into the universe.

9.6806. Eviction from Eden is Salvation..

The eviction from Eden was linked with emergence of human mind. Eden and mind are incompatible, happy idiots are deprived of any mind. Mind devolves and dies in Eden.

9.6807.

Human mind was created foolishly. Having done some foolish thing, human was evicted from Eden which made him start thinking intensively.

9.6812. He who controls questions, controls the situation.

Questions are more important than answers. Questions control answers. The right questions are awful scarce, there are not enough questions for everyone. There are extremely few questions, while answers are abundant. For every question, there are thousands and millions of answers.

9.6819. They hold their ears and close their eyes.

People deny and avoid anything they dislike. People ignore all sources of information that they dislike. People ignore information and facts that they dislike. Everything that doesn't fit their vision of the world is what people try to ignore and deny.

9.6851.

Brain needs fats, brain won't work without fats. That's why intellectual workers always have weight problems.

9.6867.

Many things are really exaggerated. while others are belittled. Rational sanity is the ability to give things their real state.

9.6883.

The world is 3-dimensional, but many people are 2 dimensional,and some of them are quite like points.

9.6884. Ghost.

Dying, a person leaves the real world. From this perspective, the people living in illusory and fantasy world can be called ghosts.

9.6908. Line and plane.

Line is an artifact of two-dimensional space, transitory stage from one-dimensional space to three-dimensional one. Evolution of one-dimensional space into three-dimensional one skips two-dimensional space. Evolution moves by leaps.

9.6909.

I've noticed that evolution moves by leaps.

9.6933. Digital obscurantism.

The problem of our time is that a person can use the Internet to find confirmation of any idiotic idea that has crossed his mind. That's why our era can be called the era of «digital obscurantism».

9.6934.

In olden days, men used to marry and have children in their early youth because life was short and it was advisable for men to have children before a war or an illness would kill them. The idea was to avoid one's lineage from being ruined. I've no idea why in our days men should get married and have children before the age of 37. Probably out of habit.

9.6942. Archived people.

They immerse unnecessary people into illusions and fantasies as a sleeping person consumes little energy and takes little space.

9.6962. Virtual food.

Restaurants will die out soon and food will become electronic. The food will be biomass with an electronic taste.

9.6969. Wasn't for bad luck, wouldn't have no luck at all.

Once a huge meteor fell on the Earth, turning it into a fiery hell. Oceans of lava covered the whole Earth, producing huge amount of gases, which become the basis of the Earth's atmosphere. Part of the Earth's mass was pushed out into space and formed the Moon which substantively contributed to the development of life on the planet, too.

9.6970. Brain overload leads to its emergency cut-off.

You can eat not more than a particular amount of information during one day. By overeating the videos and other stuff you overload your brain and cut it off in an emergency mode. Try to avoid excess information.

9.7000. Instant calculations.

Quantum method of information transmission works in one-dimensional space without time and distance. That's why such system instantly performs all calculations, it doesn't need time.

9.7018.

Acceleration is exponential, braking is too. The greater a man's strength and power, the greater his acceleration.

9.7061. Peace with yourself.

Human is a complex symbiotic system, where many living microsystems live. It's desirable to maintain peace among them, preventing rise of ones and humiliation of others, ... dividing and conquering, uniting and conquering.

9.7076.

Lie differs from truth only on the inside. They can be exactly the same on the outside.

9.7133.

One may cry: Why was I born here and not there? Why am I not the Queen of England? It's because any child is the biological clone of his parents. Essentially, he is the reincarnation of his ancestors, so he reaps their vices and virtues.

If you want to get out of the underworld, do heroic deeds and be virtuous, and then your next reincarnation (your child or your grandchild) will have a new life.

9.7139. Pure intelligence and the Beast.

The problem with many people is that they underestimate the power of the It. The power of the monster It is absolute. The subconscious controls all emotions and night dreams. All of your I want/I don't want lists, your likes and dislikes, your fears, worries, pleasures, rest, silence, joy, warmth... That's not you... It's the It. You are your inner voice, intelligence and logic. Intelligence requires the strength of mind to resist the It. Somnolent intelligence engenders monsters. Overcoming the It is impossible, but you can make arrangements with him and control him.

9.7143. The Eternal Spirit.

Man often wonders why he is here and in this body. Why not in some other body? Because you are the clone of yourself. You are the reincarnation of your ancestors. You yourself created your current life in your past lives. If you want to change something in your life, start doing it right now, and maybe you will achieve prosperity in your children and grandchildren if you turn your brains on.

9.7144.

A soul is not a solitary essence but a collective spirit. The spirit of the grain. A soul is a cluster and can be cloned through grains of intelligence. You... Your ancestors and your children are the same soul and the same spirit distributed in different bodies. A soul is a family community. If a family is united, its spirit is strong and powerful; if it is disunited, its spirit is weak. Spiritual strength determines the spirit's ability to control energy, on which the strength and the wealth of the family and kin depends.

9.7145.

Souls reproduce through spirit grains. You children are your reincarnation; they are yourself. If you have several children, all of them are your clones. However, you transferred them the grain of your soul and you need to bring them up. Caring for your own children is caring for yourself and for your own soul. Your children, your grandchildren and your great-grandchildren are your reincarnation, yourself and your eternal life. When you ask yourself several generations later "Why do I live like that?", the answer will be you today. You are responsible for yourself.

9.7146.

Everything is not contingent... God is plan and order and each of his actions is double-checked and necessary... If you see a system, you see God too. Your mission in life is to love God, that is, to serve him. Serving God brings joy.

There is the Devil, though. The Devil is contingency and chaos. The Devil means no system and no plan... Loving the Devil is not recommended and it's better to avoid him because the Devil is sufferings and pain.

9.7147. A simple truth detector.

The It is unconscious but tries to serve God. That is, if the It feels the system, it begins to serve it. By determining the systematic and recurrent nature of events, the It identifies them as a truth and begins to serve it. This is why people are obsessed with making the same mistake again and again.

9.7204. The Moon is a tourist attraction.

Nowadays few are interested in the space, except tourists. The Moon has been turned into an attraction of unheard-of joy.

9.7212.

It's only combining the incombinable that can help to achieve wholeness.

9.7226.

Inside a man's own world there can be three gods - Intelligence, the It and Ego – so there are Order, Chaos and Nothingness in the real world. Nothingness is what decides who is stronger now, God or the Devil, and it also decides which one of them will fill Nothingness up.

9.7231.

"Nothing" is a judge. Emptiness chooses itself what it will be filled with, chaos or order.

9.7234. The boundary between intelligence and illusions.

You must have heard of the fight and unity, Yin and Yang, chaos and order, good and evil, contingencies and system, unity and contradictions... What matters most is intelligence and illusions. One could say that our entire world is built on the boundary between intelligence and illusions.

9.7235. Human brain structure.

Three-level human. Atop he has mind, a little below – animal, at the base – a reptile. Once mind control becomes weaker, the human turns into an animal and sometimes even into a reptile. Human lives by mind, animals – by senses and reptiles – by instincts,

9.7236. People, cats and reptiloids.

Reptiloids are the ones who follow their instincts. A man resembles a reptilian if he is in the power of instincts. Women are usually slightly above men, since the brain of mammals is subject to feelings and emotions. Above them all are rational people who are subject to intelligence and logic rather than feelings and instincts. Formally, a woman living by feelings is closer to being human than a wild man following his instincts.

9.7239. A hamster worker.

The wheel is not only a source of joy and fun for a hamster – it is also work. The point is, the owner likes it when a hamster is running in its wheel, for this show rejoices his soul. A hamster wheel is a source of joy for its owner and the reason why he loves it. If the hamster stops bringing joy to his owner, the latter will probably have to get rid of it. A hamster wheel is a source of energy, a living electricity generator. If electricity runs out, the owner will be mad with rage.

9.7258.

Feminism is men's weapon against their most-feared enemy, women. In order for men to defeat women, women are to be turned into men.

9.7342.

Many people have distorted reality perception. He tells you: "Give me that red round thing " – meaning in reality that green square one. At all times it is advisable to clear up most thoroughly and scrupulously what exactly this individual means to say.

9.7360.

One is whole, zero is a detail. One always consists of zeros. The union of zeros turns them into one.

9.7367. The balance of the whole is when there's 7 and 3.

9.7368.

A genius man consists of love and talent. There is 1% of talent and 99% of love for the work he does.

9.7370.

One is a whole that can be composed of 3 or 7 parts.

9.7373. Lazy brain.

Human brain is lazy, the first impression defines much. Having made a decision once, further it will only explain why this decision is good and correct, while the alternative ones are wrong. The brain is easily to deceive, it may be let to fall in love with a beautiful image and later on replace it stealthily by anything at all and the brain will continue to love this anything through inertia.

9.7374.

The Universe will die when the quantity of information becomes equal to the quantity of energy. Originally, the quantity of energy in the Universe is a constant while the quantity of energy-linking information constantly grows. Originally there was energy only, later there appeared information and its quantity started to grow. What will happen if information structures all the available energy? Will the Universe shrink back into a dot?

9.7375.

Perhaps, black holes are some kind of space archivers, the mechanisms which function is to make the universe shrink back into a dot.

9.7378. An archived hamster.

A unidimensional space is a point. A two-dimensional one is a circle. A three-dimensional one is a spiral. Polar people are two-dimensional and flat, looking like hamsters doomed to eternally run in circles. However, if a hamster gets off its wheel and goes to sleep, it will turn into a point.

9.7450.

The spiral of evolution is a response to the changing facts of life. In case facts of life do not change, however, is evolution possible?

9.7463. Curiosity.

Unintelligible and new things attract attention because they may be dangerous or edible.

9.7464.

Danger, edible things and sex attract attention.

9.7515. The It (ONO), Ego and Mind.

Mind is a symbiotic entity divided into three parts, each of which is in unity and conflict with the other parts.

9.7518. Drugs.

Nanophilosophy understands drugs as all processes related to obtaining pleasure, joy and happiness. The greater pleasure is, the more it makes man dependent and addicted. What is special about feeling chronical pleasure is that feelings turn the brain off and a man loses the ability to reason. However, pleasures are not dangerous per se, but the excess thereof is. Extremes are dangerous.

9.7521. Invented happiness.

The only way to find happiness in a world without happiness is to invent it.

9.7538. A view from the window.

What matters most in any point of view is whence and where it looks.

9.7566. Movies and life.

Movie heroes die when their actors get new roles. One life ends and another one starts.

9.7577.

In fact, any person is a three-headed monster, as the It (ONO) Mind and Ego always argue in his head.

9.7590. Social suggestibility.

Women are usually very attentive to what other women like or not. To take a firm decision about whether she likes something or not, a woman usually has to see how other women respond.

9.7598. A universal female desire.

All women want the same thing. Any woman wants other women's jealousy. Nothing makes a woman's soul happy like the admiration, jealousy and glint in the eyes of other women.

9.7609.

Robots are artificial insects. Robots are strong but their brains are as big as the insects'. Evolution, however, is not static and robots will soon become reptiles... At that stage, people are likely to die out because last time dinosaurs ruled over the world for as long as 60,000,000 years.

9.7615. One consists of zeroes.

The essence of grapheme and "sorites paradox" is that the whole is substantively more than its separate parts.

9.7620.

The sense of taste is developed by uploading samples of good taste to the brain via eyes. You fill with beauty by observing beauty.

9.7628. The evolution of a point.

A point is unidimensional, a circle is two-dimensional and a spiral is three-dimensional. A hamster running round the system makes it dynamic.

9.7631.

The creation of a drawing is the destruction of perfection. The painter creates chaos by destroying the ideal perfection of a blank, white sheet of paper.

9.7635. The origin of a point.

A blank, white sheet of paper is the perfection of order. When perfection dies, it turns into chaos. A point results from the weakening of order and chaos' breaking through the canvass of reality. Further on, as perfection dies, a point turns into a new universe. Life is the movement towards death, from perfection towards chaos.

9.7650. The strong in spirit.

Nature wants to eliminate weak men, but it loves and appreciates strong ones.

9.7651. Borders of the aquarium.

Human thought can be compared to a magic goldfish, living in the aquarium. On the one hand, its mind is highly limited, on the other hand, - given the will.

9.7658.

The dark matter in the Universe is something like bad loans and dead money accumulated in the economy.

9.7668. Viruses are cause of the mind emergence.

Human is an example of viruses evolution. Viruses laid the groundwork of the human mind, putting the idea of the endless self-destructive growth in it.

9.7669. Viral nature of mind.

Mind emerged as a virus, infecting the brain of the mammal primates.

9.7689.

Living matter is inseparable from consciousness and the objective is inseparable from the subjective. If the subjective is drawn from the material, what is left is dead matter, devoid of life. The subjective breathes the spirit of life into matter. The subjective is what determines the nature of the objective but, having a close, dialectic relationship with it, it also is under the sway of the reverse relationship. Matter influences the spirit and the spirit influences matter. The stronger one influences the other. The strong in spirit can radically alter the nature of the material world around them.

9.7696. The well effect.

The future is seen much better from the past than from the present.

9.7697.

If something exists, there must be compelling reasons for that.

9.7700.

A pigopithecus is an intelligent life form on Earth that is closest to pigs.

9.7753.

The matter of existence is liquid and time is electricity.

9.7754. The solid consists of the solid and the empty consists of the empty.

The whole consists of the whole. Any detail must acquire comprehensiveness before becoming part of a whole.

9.7760.

Inside each elementary particle there is a universe. If such a particle is split, a big bang will occur and a new universe will emerge. A big bang is the process of transforming a uniform space from one dimension into another.

9.7763. Inside a cow there is an entire world.

9.7781. Dark energy is dead time.

Darkness is dead light. In the same way, dark energy is something that has no motion in it, i.e. dead time.

9.7782. Light is a predator.

Darkness is the absence of food. Light is the presence of food. Life dies if it has nothing to eat. Life is a predator.

9.7789.

Life often punishes a man out of humanistic considerations. It wants what's best for him and chases him with a stick towards the best.

9.7790.

Much of what seems bad to a man is, in fact, good. The man's incomprehension of the situation is related to his limited knowledge about external circumstances and the system's system-wide motion.

9.7803.

Not that NOTHINGNESS is bad, but it is not good either. Nothingness is a place for a new life, but nobody knows in advance what kind of life will originate there.

9.7804. Purpose of mind.

I gave you many various answers. You should choose the most appropriate to your current conditions yourself. That's what you need it for.

9.7810. A life system.

Science does not deny the idea of God - only fools deny it. God is truth and the mind. If a person has no mind, he cannot see God.

9.7811.

Religion is a symbol of God rather than God himself. Don't mix up these two notions. There can be a lot of symbols, but there is always just one God.

9.7813.

Human suffering is part of the divine intention and is related to the expulsion from the human mind of the damned, stupidity, vices, weakness, idealism and other signs of idiocy. Rationalism is intelligence and God is Intelligence. Suffering makes people rational.

9.7816. Often 7.

The reasoning system where one question means one right answer turns a person into a slave, depriving him of the freedom of thought. In the real world, any question has at least three right answers.

9.7817. An old biblical story and yeast.

Meteorites brought intelligence to the Earth from space. Intelligence is a fungus parasitic upon the brain of mammals. Having come from space as spores, the fungus developed into mycelia and infected fruit trees. One day, a monkey overate apples and got infected too. You know what happened next.

9.7850. How economic bubbles work.

A group of huge investors form a starting pooled fund, whose finances are spent on advertisement, hype creation and control of the "bubble stock rates". Drastic controlled rate growth, as well as the stories about those who got incredibly wealthy due to it, cause the «Sancho Panza» in people, turning their mind off. Itchy for gain, thousands of people invest hundreds of dollars (not to risk very much). Next it's quite possible to pay shares to the system organizers out of these money and burst the bubble. In this case, the system organizers, the big players, will earn money and millions of the secondary small players will lose a few hundred dollars. But as the sums are not huge, nobody will suffer much. But there is a second option: it's possible to pay fist small shares with a promised income and provoke people to invest tens of thousand dollars and only then bankrupt the system. The second option is rather cruel and inhumane. To lose a few hundred dollars is one thing and to lose tens of thousands is quite a different one.

9.7851. Cryptocurrency.

Blockchain-based financial instruments. An electronic illusion, that is not asset-based and not protected by law. A kind of classic speculative bubble of the «Dutch tulips»-type in the light of the current reality. In essence, it's gambling where the winners are those who managed to leave it before the bubble burst.

9.7868.

It seems that the black color is nothing, that is, it is non-existent. Then the question arises as to why honey will go bad if honey is mixed with tar.

9.7869.

I deny the stronger and the weaker sexes. I distinguish only between the weak and strong in mind. The weaker sex are those who are weak in spirit and in mind. Both men and women can be the weaker sex.

9.7877.

Regularities can be direct and contingent. Contingencies have to do with elliptic motion rather than direct motion.

9.7878.

A hamster running in its wheel is an essence trapped in a time loop. A beam of time consists of a multitude of parallel circles, and one hamster runs in each of them.

9.7882. The material and the immaterial.

The objective is the effect and the subjective is its cause.

9.7883. One is the consequence of zero.

One is the effect and zero is its cause. Darkness is the cause of light, chaos is that of order and evil is that of good.

9.7885. Intelligence is not a virus but a fungus.

A human community looks very much like a mycelium.

9.7892. Divine rationalism.

God is the mind. The one who serves the mind serves God.

9.7897. DNA principle

The problem of a hamster in a wheel is that it resets to zero at every round. If the hamster used the DNA principle and saved 50% of its past experience, it could get to a new level.

9.7902. Overcome with emotions.

Women are called the weaker sex not because they are

weaker physically, but because they are weaker in spirit. Women find it hard to keep control of their fears and emotions, anger, emotional pain, thirst for pleasure, love, etc.

9.7909. Three regulatory systems.

The nervous system, the endocrine system and the immune system. Brain electrical impulses, hormones and cytokines. All the three systems work in the single harmony and are interrelated. The key is that the mind, controlling the nervous system, can affect both the endocrine and immune systems. In other words, way of thinking defines the health of the organism.

9.7910. Allergy to other people's opinion.

The immune system defines the organism's reaction to everything that's foreign to it, including stranger people, foreign thoughts, foreign influence. All this affects hormones and such things as laziness, sleepiness, attentiveness, anger and so on.

To suppress the immune reaction, the brain needs to strictly control its emotions, otherwise emotions send a signal to the endocrine and immune systems, hormones and cytokines are released into the blood... And the mind loses control over the situation.

9.7911. Wooden man.

Indeed, jamming nervous system by sedatives or dopamine is efficient and it solves many problems of the immune and endocrine systems, but... Sleeping brain loses a lot of its useful capacities and the person starts to resemble a plant in some way. Sensible control of emotions would be much more efficient than banal brain shutdown.

9.7913. A conditioned reflex.

An unconditioned reflex is what a man is given from his

birth, that is, his genetic make-up and parents. However, you will not get far with unconditioned reflexes, so there are conditioned reflexes too. These are the properties of the mind that you can develop on your own: philosophy, spiritual strength, mental strength, upbringing and so on.

9.7926. Reminder of a combat virus.

You need novelty and uniqueness to prevent the immune system from identifying you as an enemy at first. It will give you some time. But you should act very quickly. Very soon you will be identified and destroyed, if you don't manage to multiply and take over the base.

9.7928. Masters of money and pleasures.

The modern society development condemns the institution of slavery. A person can't belong to a person. We stand slaves of money and slaves of pleasures, that's enough.

9.7937.

Beauty is a source of joy. Beauty is a pure pleasure, a boost of vigor and energy.

9.7941. A square is harmony.

A square is an object on which seven perspectives are possible.

9.7942.

First, evil is to be annihilated and, when it does not work, you can try using a terrible weapon such as love.

9.7944. Because it should be so!

Much of what should not exist exists. Why?

9.7945. Beauty is logic, beauty is truth and beauty is harmony.

In this world, beauty is the only thing that can be trusted because beauty is this world's god.

9.7958. They cry as they are blind.

Mind is an ability to see. The crazy are blind to the truth, to regularities, to logic, to beauty, to money, to wonder.

9.7959. Half measures do not fix anything.

I noticed that any problem can be resolved drastically.

9.7960.

Rationality is an ability to see beauty. The more complex and unique beauty a person can see, the more sensible he is.

9.7963. Friend or foe.

A foreign thought should cheat the immune system to get into human brain. It should pretend to be your own thought or a neutral one. Foreign hostile thoughts are killed by the brain immune system at once.

9.7974.

The point in the simultaneity of the states of existence is that this world is simultaneously unidimensional and multidimensional. Everything is simultaneously unidimensional and multidimensional down here. Unidimensional and three-dimensional, two-dimensional and four-dimensional, unidimensional and seven-dimensional, two-dimensional and eight-dimensional.

9.7975. There are usually 3 to 7 reasons.

Everything always has at least two reasons: one reflects form and the other one content.

9.7976. Brain gyri are like abs,

The people reading books for pleasure usually suffer from

brain obesity. One should read the books as he does sports, forcing himself to train and mobilizing his brain.

9.7979. I DON'T WANT.

Ready-made thinking is when everything that isn't like with the others or isn't as per usual is rejected and causes discomfort.

9.7988. Dynamic 3D truth.

The essence of the fourth dimension is that the first three truths are constantly changing.

9.7994.

Human is unable to acquire the external information as a whole. The brain immune system prevents human from acquiring the foreign. External information gets to the human brain in the form of grains. And what will germinate and grow, becomes his own personal information, which the immune system identifies as its own.

9.7995. Misperception of reality.

Robots perceive information objectively and people do it subjectively. It can be said that human communicates with the world by a subject-oriented interface. Human is almost unable to observe the objective world. Bad memory distinguishes people from the robots very favorably. Absolute memory makes people extremely unhappy and foolish, they have no reasons to second-guess the information. The main resources of the human brain are aimed at second-guessing of the external information.

9.7996.

Later they defeated death, but this is what destroyed them. Having gone mad with infinite boredom, they self-destroyed very quickly.

THE THEORY OF EXISTENCE

9.7997.

Love is a mechanism for concentrating pleasure. What a man loves brings him as much pleasure as possible. Animals are constrained in terms of love, but humans are capable of loving anything.

9.7998.

Contingencies break rules. Rules are God and contingencies are the Devil. Since God does not love the Devil, contingencies are unpredictable and quick. The Devil wants to go unnoticed, so he prefers acting unpredictably and quickly.

9.7999. The weapon against boredom.

Man is God's toy. In creating man, God was thinking about how to kill his boredom.

9.8002.

Mind is the breed of the subjective, mind strives to destroy the objective world.

9.8004. Human identity structure.

The four essences of the brain: Mind, Ego, the inner IT, the social IT (outside the human). Mind and Ego are in the neocortex. The IT is the reptile brain (the instincts) and the limbic brain (feelings and emotions). The social IT is in the socium surrounding the human.

9.8012. The placebo effect.

Most problems and difficulties disappear by themselves, incapable of resisting the immune system of the healthy nature of things. The immunity of existence will cure by itself all devilish manifestations of the contingency of chaos, if one does not interfere with it or ruin everything.

9.8013. Ignore the system's contingent fluctuations.

It's better to ignore the Devil. Chaos' contingencies are very weak and the immunity of existence kills them very quickly if they are not interfered with.

9.8016. Vampires.

Contingencies are weak and disappear by themselves unless you start fighting against them. When fighting, they find the strength to defeat anything.

9.8020.

Fuel and an oxidizer are needed to produce energy. The fusion of light and darkness and of order and chaos are an endless source of energy in this world.

9.8023. The chain's thinnest link.

Curiously, life on Earth depends on just one bacteria. If something destroys cyanobacteria, all life will perish.

9.8026. A catcher of contingencies.

Chaos' contingencies are almost instantaneous. Not because order kills them, but because they want freedom and try to run away from it. Order aims to catch a contingency and to make it its source of energy by turning it into a regularity.

9.8028.

Uniqueness is the air of things perfect.

9.8029. A systemic element.

The male bee only seems useless. In reality, the beehive will die without him.

9.8036. Mind is a predator.

Mind is the tool to get energy, to get food.

9.8037.

God is a phenomenon of the objective world because God is order, pragmatism, truth and intelligence. The Devil is the opposite. He is a phenomenon of the subjective world, deception, illusions, rubbish and idealism.

9.8038. A contingency is something unexpected.

The anticipation of a contingency kills it and engenders a new, unexpected contingency.

9.8041.

Tactics is the art of achieving the goal. Strategy is the art of setting right goals.

9.8042.

Everyone has almost the same brain, but the software is different with everyone.

9.8045.

Taking a wrong decision quickly is better than getting stuck, unable to choose anything. What's more, taking a decision is less important than persevering in sticking to it.

9.8058.

At a cat exhibition, I was staring at people with great interest. They were all different from each other, but they looked very much like their cats.

9.8059.

You are looking for extraterrestrial intelligent life in the wrong place. You should look for it on the Internet. Rumor has it, an entire population of reptiloids live there openly.

9.8061.

Many things happen just because somebody wanted it very much. You search for a deep sense of something happening a particular way, but the answer is simple - somebody wanted it to be so. Somebody loved his desire very much, so he made every effort to make it come true.

9.8063.

A hamster running in its wheel is the symbol and the metaphor of a perpetual motion machine.

9.8066.

Critical thinking is the ability to quickly assess if there is any profit in something or not.

9.8069. A blessed vegetable.

Sects feed huge portions of happiness to their followers, which is addictive and turns people into addicts. Blessed vegetables can no longer reason or live independently.

9.8070.

Quantum philosophy perceives the subjective and the objective as a single harmonious essence. This is a specific, harmonious view of the world that rejects the eventuality of the separate existence of materialism and idealism.

9.8071.

The evolution of thought from zero to one is the development from the subjective to the objective, from an idea to its implementation into reality. The real world is objective: when you are working on a building or a car, there should be nothing indefinite or contingent in them.

9.8072.

Subjectivity is part of a growing thought. For a thought to be materialized, however, it should become 100% objective.

9.8073.

The Universe is objective. The point from which it originated is subjective.

9.8094.

By fighting emerging threats to itself, immunity is continuously educating itself and generating new conditioned reflexes in itself. In the same way, brains should confront and overcome a problem in order to acquire new knowledge. This will produce a conditioned reflex necessary to fight this kind of threats. The larger the number of threats confronted by brains, the better the latter's immunity becomes.

9.8120.

They love all sorts of secret formula, well, if you love someone, give them what they want.

9.8122. Quantum supercomputers.

The genuine non-linear strategic thinking, based on such a tool as intuition is inherent to the quantum way of thinking. This way of thinking is characterized by the simultaneous analysis of the incongruous factors.

Creating a single multifactor model of the situation and evaluating it not by the usual consequent-linear way, but by parallel-quantum way under the simultaneous analysis of all the system conditions.

9.8124. Software and hardware.

What does a person need to start thinking in the quantum way? Firstly, he needs brain, and secondly, the right philosophy.

9.8125.

The quantum way of thinking means congruence of the incongruous: the rational way of thinking - logic and subjective-probabilistic synthesis - of the quantum physics.

9.8128.

Syntalism is a materialist philosophy based on cybernetics, pragmatism and logic. Syntalism is rooted in rationalism, stoicism and even quantum physics. Quantum philosophy considers the subjective part of reality from mechanistic positions. The subjective is the reflection of the objective, it's a whole unity, two different sides of the same coin.

9.8137. A reference man.

God is the golden mean and the Devil is extremes. God is a system and the Devil is chaos. Sinners are full of extremes, but the closer a man gets to God, the more normal they become. Most people prefer these or those extremes. The closer to the edge, the fuller a man's life is of the Devil and contingencies. The closer one gets to the golden mean, the more there is God, energy and systemic Miracle.

9.8139.

Extremes are very much alike: one extreme easily flows into any other opposite extreme. There are usually three extremes in a three-dimensional life. In a four-dimensional life, there are four of them, but they are constantly changing places or inside themselves.

9.8149. A swarm of points.

Points are weak. In concentrating your efforts into one point, you will defeat this point very quickly. However, sometimes points form a swarm and there is no escape from them then.

9.8176.

It was observed that many women would like to earn more money than men, but they wouldn't like to be in a relationship with men who earn less money than themselves. It follows that if these women's first wish comes true, the humankind will become extinct.

9.8180.

God created an ideal world, but he did not create an ideal man. Now this non-ideal man is forced to constantly remake the ideal world for himself.

9.8186.

Evolution is always a reasonable compromise. The strong want to eat the weak, but the weak doesn't want to be eaten by anybody. Moreover, if the strong eats all the weak, the very next moment he will become the weak link himself and die of starvation.

9.8187.

If God created man in his own image, that's really horrible.

9.8188.

Of course, the world is not ideal, but nobody has not yet managed to create anything better in the past few billions of years.

9.8206.

What prevents a man from reasoning independently is cognitive dissonance.

9.8207. The source of all evil.

According to Freud, tough childhood and parents are responsible for all human troubles. Modern psychology, however, proved that only Cognitive Dissonance is to be blamed for everything, actually.

Cognitive dissonance is, essentially, the equivalent of the Freudian It, the biblical Devil and Islamic Nafs... It is also the paradigm of the weakness of the human spirit that cannot resist his feelings and emotions on his own.

9.8209. The spirit's weakness.

Modern psychologists revealed that cognitive dissonance is the source of all human troubles and man himself, as usual, is not to be blamed.

9.8210.

The commonplace "I don't want" results from cognitive dissonance. Everything causing cognitive dissonance is not usually really wanted.

9.8221.

People full of extremes are form people or empty people devoid of content.

9.8222.

Extremes are form. There is content inside from – it is moderation. It often happens, however, that a man is empty inside: there is no golden mean in him and there is only the shell of extremes.

9.8248.

Regarding the expansion of the boundaries of intelligence, one could say that beyond the boundaries of intelligence is its absence, i.e. stupidity.

9.8275.

The search for good spirits is the driving force of evolution.

9.8276.

Pain is the driving force of evolution: when fleeing from pain, a man runs to the future at breakneck speed.

9.8296.

The scope of a personality is very dependent on his access to television. The greater the personality, the more space it takes up on TV.

9.8308.

Only 1% of a mountain is its top. In terms of energy, it is very little, so you mandatorily need the remaining 99%.

9.8309.

Only 1% of a mountain is its top. In the same way, only 1% of a genius is his talent.

9.8324. Water.

Time isn't money, time is one of the energy states, that can be transformed into the money under some conditions, or not to be transferred. And the rate is always changing and is very dependent on the external conditions. Time is goods, not money.

9.8325.

Paradise abounds with illegal damned ones. Basically, the damned are in charge of all dirty work in paradise.

9.8335.

There are addition and multiplication, and then there is a synergetic operation. It's when a whole is being composed of pieces.

9.8336.

Love is the feeling of comprehensiveness. If you have no love, you will never achieve comprehensiveness.

9.8344. The Holy Trinity.

There are actors, directors and script writers, and then there are three-in-one people. They are the noble people playing life according to their own scenario.

9.8355.

One day, the planet Earth got sick and ran a fever. Glaciers on the poles started thawing, the Earth got all sweaty and the weather broke, which did not kill people, though. What killed people were the antibiotics that the physician pre-scribed to the patient.

9.8369.

The Devil is the supreme demon, and the damned one is the demon's slave, that is, a cursed man who sold his soul to the devil. Demons can be different and each sin and vice has its demon. As for the Devil (also known as the Satan or Iblis), it means the unity and comprehensiveness of vice, the union of all other demons.

9.8379. Free people.

Slavery has not been abolished. Slaves of money and slaves of pleasures have been, are, and will be forever.

9.8382.

We don't have slavery, but we have slaves of money, slaves of pleasures, slaves of vices, slaves of sins. The damned are the devil's slaves, who have sold their souls to the devil and are doomed to torments and suffering in hell.

9.8394.

Life forms can be different. As an example, there are such dumb ones that you don't even feel like having faith in them.

9.8395. When it really sucks.

I wouldn't idealize extraterrestrial intelligent life forms: most likely, they are as dumb as humans, and maybe even dumber than that.

9.8396. Another big problem.

Mystic life forms such as angels, the damned ones and some other gods... are metaphors and symbols. However, if we suddenly decided to imagine them as really existing life forms, this would do away with their mysticism, turning them into intelligent life forms competing with us, humans.

9.8400.

The wisdom of millionaires can't help mere mortals like us. It's easier for a lion to be the monarch of all beasts, but we're mostly interested how a rabbit can be the monarch.

9.8401. The injustice of existence.

We know a lot about wolves in sheep's clothing and almost nothing about sheep in wolves' clothing.

9.8406. Books are monotheism and movies are paganism.

A book's God is its author, but there are a lot of gods in movies: a script writer, a director and quite a few other minor gods and goddesses.

9.8408.

You are often dissatisfied with the god of our world. You shouldn't be, though. You are really lucky that our god is the ideal of normality. I've seen many other worlds in books and movies... and, you know, I'm very glad we've been so lucky with our script writer.

9.8427. Programming conditioned reflexes.

Feeling jealousy or overwhelming desire?.. Know that they want to cause excessive salivation in your mouth, just like Pavlov's dog, and turn you into a zombie.

9.8430.

A universe originates from a point and a man from an egg cell and a sperm cell. The question is, if the point was an egg cell, what was a sperm cell?

9.8434.

The same problem always have dozens and even hundreds of solutions.

9.8435.

Any problem is a task that has an infinite quantity of solutions.

9.8464.

A damned one is an aggressive, mystical life form – an intelligent parasite that damages the human brain and takes control over the human personality, turning him into the It's slave.

9.8465.

Ironically, the desire to be exceptionally unique makes them extremely similar to one another.

9.8469.

The Internet abounds with unreal people. Nobody knows if there are any real ones.

9.8487.

The less there is mysticism, obscurantism and talks with aspiration in mysticism, the better. The closer mysticism is to logic and pragmatics, the better and more effectively it works.

9.8513.

They like to emotionize rather than think. Emotionizing is a process of deriving pleasure, instead of meaning, from information.

9.8519. Philosophy is the basis of human identity.

Neither race, sex, age, nor even education define human identity, but it's his philosophy that does.

9.8542. God was an engineer.

There is engineer thinking and there is its absence. It's like there is mind and there is none. Rationality is the ability to create thought.

9.8546. Extremes are the form of content.

Extremes are a form controlling the golden mean – content.

9.8547.

Humans are doomed to die out. The only way humans can avoid dying out is to stop being humans and become...............

9.8548.

People will surely inevitably die out, but mind will live forever.

9.8549. Median chaos.

Extremes have strong convictions and are similar to form. Chaos and uncertainty are raging in the middle of this system.

9.8551.

The characteristic of rationality is the ability to take the useful out of useless, separating the wheat from the chuff.

9.8554.

Life can be compared to a chess game, but that is a very weird chess game. Imagine a multidimensional chessboard with hundreds of games going on it simultaneously and all players constantly disturbing and interfering with one an-other.

9.8558. Separator.

The art of rationality is the ability to take the fly out of the barrel of ointment.

9.8563. Mission of mind.

The unconscious should be controlled. The unconscious should be made conscious.

9.8571.

The physics of human relationships is based on the laws of gravity.

9.8572.

All have the same problems but different solutions.

9.8590. A positron.

The essence of truth is related to a phenomenon known as annihilation. Truth and lie are destroyed by coming into contact, but electron-positron pairs come to replace them.

9.8609.

A normal person combines three types of the intelligence - social intelligence, emotional intelligence and logical intel-ligence.

9.8636.

What is needed is the balance of the incompatible, other-wise nothing good will come of it.

9.8637. The iron planet.

The Earth resembles a Spinning Top with a liquid iron nucleus inside. The axis of rotation of the spinning top is constantly changing, thus modifying the iron planet's climate.

9.8638. Seize the day.

The presence of intelligent life on the planet Earth looks very much like a diabolical accident. The climate that produced intelligent life is but a second in the history of the planet Earth. A second later, the degree of the Earth's axis of rotation would change and its orbit would no longer be a circle but an ellipse... The climate would change drastically too, making the existence of intelligent life on the planet utterly unbearable.

9.8645.

Socialization is very important for the even-toed. The bigger crowd the dominant individuals have gathered around, the safer the herd center will be. Predators usually catch lost sheep or those by the sides.

9.8646.

Socialization is good. They walk by bunches and groups, smile and laugh synchronically. They are all as one, from afar these small flocks of birds seem one big animal which predators are not so eager to deal with.

9.8672.

Dreaming is a basic form of planning.

9.8674. An augur.

Fortune-telling is less related to mysticism than to mathematics, physics, the theory of probability, history, sociology, psychology and stuff like that.

9.8677. A vicious circle.

Thoughts determine consciousness, consciousness determines reality and reality determines thoughts.

9.8688.

Having faith in nonsense makes sense only if it brings profit or, at least, pleasure.

9.8689.

Rationality is the ability to get benefit. Mind is the tool made for getting benefit.

9.8693. Search for thought is the search for silence.

The silence of thoughts is necessary for real thinking. It's necessary to clear your head from the unnecessary thoughts and even force yourself to keep silence.

9.8697. The foundation of truth.

A number of verified people can be trusted. Hegel is always right, Nietzsche is always right, Freud is always right... If you think they were wrong, you are in the power of illusion.

9.8720.

The point of the evolution of a population is that a bad example sets a good one and now everybody knows what is not to be done.

9.8736.

If you do not understand something... ask – do not suffer ahead of time. Actually, humanity has not yet come up with things that are impossible or too hard to understand.

9.8738.

Space time is said to be homogeneous and the time vector the same. Imagine that it is not so, though. Imagine that the time vector is contingent and only the sum of all those contingencies produces their unity. Of course, this river flows in one direction, but the fish in it can swim anywhere.

9.8739.

Time is order, but any order seeks chaos.

9.8740.

One expands, first, into three, then, into seven in a two-dimensional space. One axis of measurement is time and another one is energy in a two-dimensional space.

9.8741.

The amount of energy in the unit of time is the entity we call information. That is, energy exists ab initio and information is time and its derivative. The amount of information increases in a system over time.

9.8742.

The amount of energy in a system is the invariable relative to the amount in the time system. The system's expansion in time also increases the amount of energy in it.

9.8743.

Time is the speed with which the entities in a system move in relation to one another and to the clock rate. In such a system, the clock rate is the invariable of the speed of light.

9.8744.

Energy is the entity that comprises relationships related to the size of space and its mass in relation to the reference state of one. In fact, it is a charge from zero to one per square measure.

9.8745.

The Universe is, basically, a quantum computer system in which zero and one with a variable charge level encode information. The clock rate in such a system is the variable dependent on the speed with respect to the speed of time.

9.8746.

The expansion of the universe is the process of creating information. The more the clock rate, the faster the speed of the universe's expansion and the more information it contains. As the system grows, the dimensionality of space increases – one-, three-, seven-dimensional space and so on.

9.8747.

The endlessness of space has to do with its multidimensionality. The boundlessness of the universe has to do with the stepped structure of measurements.

9.8751.

Every point has at least three levels of freedom, then everything depends on personal impudence.

9.8752.

There is minimum distance and no maximum distance. Why this injustice?

9.8753.

Three-dimensionality of the space is linked with three levels of freedom of the most primitive objects... More complicated objects have six, seven and more levels of freedom that shows that the bigger the object is, the more dimensionality of its space is.

9.8755.

The possibility of rotation doubles the dimensionality of a three-dimensional space, and the seventh dimension is the internal characteristic of the sphere of rotation.

9.8759.

Basically, one and the same impetus can cause three diametrically opposite reactions. Three impetuses aimed at the same reaction should be used to reduced the system uncertainty.

9.8760.

The psychological influence of stimuli on response should be examined within the framework of Einstein's theory of probability. As an example, the properties of stimuli such as speed, mass, force and quantity should be taken into consideration.

9.8762.

Boundlessness and endlessness are different. Boundlessness is a loop and endlessness is a myth.

9.8764. A new dimension.

To escape from an endless circle, expanding the boundaries of consciousness is not sufficient – what is to be expanded is its dimensionality.

9.8766. Three is the first level of freedom.

The minimal level of freedom starts in the three-dimensional space, when there are three levels of freedom. One-dimensional and two- dimensional people have only one or two levels of freedom, that prevents them from getting out of the endless circle and reaching the first level of freedom.

9.8768.

Uni- and two-dimensional entities have no volume and, consequently, no content. Such people are but form, devoid of wholeness and even of emptiness.

9.8772.

A three-dimensional sphere within two-dimensional space

is an example of the curvature of space. Curved space is stretched over the sphere and, at the same time, is devoid of the third dimension. The curvature of space emerges from the interaction between two-dimensional and three-dimensional space and results from the distortion of two-dimensional space depending on the radius of three-dimensional space.

9.8773.

Unidimensional space is a point, two-dimensional space is a plane and three-dimensional space is a sphere. The expanding universe is four-dimensional space.

9.8776. The curvature of time.

We know that is a two-dimensional space is stretched over a three-dimensional space in which the former will become form and the latter will become content, the two-dimensional space will be boundless because it will have no bounds. Boundlessness, however, is not endlessness because the area of such a space will remain a finite invariable.

That is, if a two-dimensional entity is given a three-dimensional form, it will be boundless but not endless. Correspondingly, if a three-dimensional space is given a four-dimensional form, it will also be boundless. Consequently, the boundlessness of our universe is due to the fact that it is a three-dimensional entity in the form of a four-dimensional one. It also means that a four-dimensional curvature is inherent in our three-dimensional universe. Considering that time is the fourth dimension in our universe, time in our universe can be said to be non-linear and curved.

9.8779.

The Universe is a three-dimensional dynamic sphere in which the fourth pseudo-dimension is a pseudo-invariable, or time.

9.8780.

Time is the pseudo-invariable of a three-dimensional space. In a four-dimensional space, time has many degrees of freedom and only one degree in a three-dimensional one.

9.8788.

A two-dimensional plane is boundless and, consequently, looped within a three-dimensional sphere's space.

9.8790.

The smaller a time span, the easier it is to predict the future. The closer the future, the more visible it is.

9.8791.

Any uncontrolled process seeks chaos. A system can be said to strive from order for chaos and from one for zero. The normal state of a system is its destruction. The fact that a system is moving from one to zero suggests that the universe's existence is finite.

9.8792.

The speed of time gradually fades, time used to move faster in the past. In the future it will become even slower until it stops at all. The universe will die the moment time will stop.

9.8793.

The speed of time is a value reciprocal of the speed of the universe's disintegration.

9.8794.

The disintegration rate of the universe goes exponentially, the older the universe is the higher the rate gets, besides finally there will be a slumping growth of entropy.

9.8795.

The higher the disintegration rate of the universe is, the slower the speed of time. The higher the speed is, the slower time goes.

9.8796.

Dear friends, all texts in "Variothoughts" are written from the perspective of quantum philosophy and may either coincide or don't coincide with viewpoints of other science disciplines.

9.8797.

The problem with predicting the future is that the future itself does not exist yet, but there is a multitude of its varieties, each of which wants, like a sperm cell, to fertilize an egg cell and become real. The best one will win, but this process is 50% contingent.

9.8799.

As the dimensionality of space increases, its area increases by the number of the primary system's elementary particles and the invariable of distance between them. In a seven-dimensional universe, a three-dimensional one looks like an elementary particle.

9.8800. The speed of light is not an invariable.

It can be assumed that the speed of light is related to the speed of the universe's disintegration and changes in proportion to it.

9.8801. Virtual time travels.

They will soon create powerful computers that will be able to calculate the future up to the last quantum. That's not the most exciting part of it, though. The same computer system will calculate the past and will create its virtual model, thus making real virtual travels to the past.

9.8802.

When traveling in time, you should know that there are as many variations of the past as those of the future.

9.8803.

If future can be predicted, a model of the past can be restored too.

9.8804.

They consider mathematics to be very complicated and, meanwhile, they see themselves as life experts. That's just ridiculous. Mathematical models are a monstrous simplification of real life. Mathematics is much simpler that reality. They seek to understand complicated things without understanding simple ones.

9.8805.

The Universe is God who created himself. So anyone, who was initially a nobody, can – by means of personal growth – creep out of his point and become a rather worthy person.

9.8806. The circle of thought.

There is nothing but philosophy in mathematics.
There is nothing but mathematics in physics.
There is nothing but physics in chemistry.
There is nothing but chemistry in biology.
There is nothing but biology in psychology.
There is nothing but psychology in philosophy.

9.8807. Drinking milk from the mirror-world is prohibited.

The mirror-world is as different from reality as antimatter from usual matter. In touching each other, matter and antimatter get annihilated.

9.8809.

Matter and antimatter get annihilated, indeed, in touching each other. What happens to information and anti-information if they get mixed?

9.8810.

There is matter and antimatter, there are particles and anti-particles, there is energy and dark energy... And what about information and anti-information? Maybe they are truth and lie, God and the Devil?

9.8811. A source of energy.

If we drew an analogy between God and the Devil as being matter and antimatter respectively, we could suggest that the two of them prefer avoiding each other because their collision inevitably ends in mutual destruction and a powerful burst of energy.

9.8814.

Everybody loves self-confident people... Self-confidence is faith. Faith is power... Faith is love.

9.8824.

There are many good people and few wonderful ones. 20% of effort is necessary to do well and to do excellently, a person should do his best.

9.8831. The principle of water.

They do what is profitable for them and it would be strange if it was otherwise.

9.8832.

Man is a liquid whose shape is defined by external circumstances.

9.8840. Variable capacity.

The capacity of quantum computing systems changes according to the object of calculation.

9.8845. Be useful.

Depression is the sign of uselessness. Depression is the mechanism of self-destruction. Nature thinks it does not need useless people and, thus, wants to destroy them with their own hands.

9.8847. A person X.

Zeros, ones, threes, fives etc are good but the real king of scene is a person X who can take any meaning depending on present situation.

9.8848.

The most valuable card in the pack is joker. Different from others, joker is not scared of unlucky circumstances.

9.8849.

Joker is the symbol of quantum philosophy. X can be anybody depending on the situation.

9.8851. Essence of X.

They say a person is a zero or a one, or some other number. It's not true. The real person is always X, as a person can actually be whatever he wants.

9.8853.

The biggest – and the smallest – digit in the universe is X.

9.8854.

X is the ideal entity that can achieve wholeness with anyone.

9.8867. A pawn.

Aces and kings are all very well, but the God of the play is the Joker.

9.8881. A paradox of socialization.

Generally, women are more social than men, but their capability to be friends and to form coherent groups bound by common goals is much weaker than that of men.

9.8891. Qubit.

X is an endlessly small unit and, at the same time, an endlessly big one. X is a quantum number that combines into endlessness. X is the simultaneous matrix of all its states, whose value depends on the current external situation. X is truth.

9.8892. Man is the derivative of Y with respect to X.

The two major essences of the universe can be identified. The essence X is the array of all internal states whose value depends on external circumstances. The essence X is Content. The essence Y is the array of all external circumstances. The essence Y is Form. The essence Y is what is inside and the essence Y is what is outside.

9.8893.

X reminds me of a balloon: you can inflate a balloon and deflate it too. When in a bad mood, the balloon deflates and, when in a good mood, it inflates.

9.8894.

Angels live up in the clouds – clouds are dreams and plans. The damned live underground because they have no constructive plans.

9.8896.

Surprising things are what we cannot even imagine.

9.8899. Energy is running out.

The amount of energy in the universe is said to be an invariable. At the same time, the universe is getting cold and the universe is disintegrating.

9.8901. Why does nothing flow into us?

The amount of energy in the universe is constantly decreasing. Black holes are the points of leakage of energy from the system through which our energy flows to other universes.

9.8902.

Energy means being in the right mood, absence of energy means absence of mood for doing anything.

9.8914.

Problems usually arise in a very untimely manner. This is due to the fact that predators seek a weakness before attacking. They expressly wait for you to weaken and then attack... On the contrary, they retreat, having felt someone stronger.

9.8917.

Time is the speed with which the universe grows cold.
... when the universe grows cold, energy will run out and time will be up.

9.8929.

Surprise is said to last nine days. That's the maximum. In practice, it can be seven days, three days and one day... Or even nine minutes.

9.8931.

Man is like a mosquito but he cannot bite even through the Earth's crust… He is just sipping some petroleum on its top, that's it.

9.8987.

A hamster wheel is an example of a perpetual motion machine.

9.8996. A running man.

The point of the human population as a virus is that it develops much faster than the immune system of the planet Earth can respond to it.

9.9007.

Of every living thing of all flesh, there are not two, but three sorts. There are three sorts of fools, three sorts of intelligent ones, three sorts of women, three sorts of men and so on.

9.9012. Mind games.

One day, we designed a trial self-learning artificial intelligence system that can simulate a game world inside itself. This self-learned game intelligence developed by playing with itself inside its virtual world. Fifteen billion years of internal game time had not elapsed when the SYSTEM managed to create an entire universe and a subsidiary intelligence inside itself. These creatures were called Humans, I guess.

9.9020. A diabolical coincidence.

IT is information technology. The It is the subconscious according to Freud and the devil according to the Bible.

9.9021. Bad structured information system.

Mind is regularities and order. Absence of mind, i.e idiocy, is chaos and the breed of contingency.

9.9041.

Freedom is when the body is not interacting with other bodies. The more such interactions, the less freedoms.

9.9042. What does not exist yet.

The speed of light is related to the clock frequency of time in the universe. In theory, you can move faster than the speed of light, but then you will outrun time and will end up in the future, that is, in nothingness.

9.9043.

Evil is useful, since one evil protects you from another evil. Nature abhors a vacuum, and you will have time to separate the wheat from the chaff before crop harvesting.

9.9051.

Evil is nothing. It is non-existent. It is zero. He who creates ZERO multiplies it by all the good things he has done in his life so far.

9.9053.

According to the Old Testament, people sunk in vice and sin are pigs. That is why Islam and Judaism condemn pork and incite believers not to defile themselves by touching sin and vice.

9.9060.

Philosophy is a vaccine against stupidity.

9.9086.

Understatement allows spectators to think up. A thought-up idea is your own idea and, consequently, a familiar and right one.

9.9090.

Intuition is needed to avoid extremes such as shortage and excess.

9.9094.

Theory defines rules in the boundaries of which practice works. Theory means power and mastership, while practice is its work force.

9.9098. The universal language.

Knowing how to draw is like knowing one more language. A drawing is the universal language that is easy to understand by everyone.

9.9099.

The mechanism that provides the process of forgetting in the human brain is very similar to artificial power limit, preinstalled at modern car engines. These limits can be removed, but what will be the consequences?

9.9100.

Bad memory distinguishes intelligent computers from intelligent ones.

9.9104. There's nothing contiguous about human.

The matter is that human is unable to generate real contingency deliberately. Contingencies created by the human, are pseudo-contiguous and, in fact, very logical.

9.9105. A contingency likes to disguise itself as a regularity.

People do not believe in contingent repetitions. However, there are many more contingent repetitions than a man can even imagine.

9.9106.

The future is what is not and has never been.

THE THEORY OF EXISTENCE

9.9110.

Truth is what is hidden in lie. Rationality is the ability to separate truth from lie, the wheat from the chaff and honey from tar.

9.9114.

Mind is external, so it's highly dependent on the place of living and the people around.

9.9115. Two plus one.

There are two program complexes in the human: the subconscious IT and mind, both essences are external, but competing. Both essences were generated by the external world and are controlled by it. The subconscious IT emerged earlier, mind - much later. Only human EGO, that these external two competing systems are trying to control, is internal.

9.9117. Soul is EGO.

The voice inside you, is that you? It's the whispering voice of either an angel or the devil. Or maybe you are your silent IT, responsible for fear and pleasures.

9.9118. Emote less.

Emotions consume a great amount of energy, depriving the rest of the brain of energy, thus turning mind and intelligence off.

9.9119.

One of the greatest human needs is love... Self-love. A man really needs to love himself and be proud of himself. A man devoid of his own love is dejected and unhappy.

9.9121.

Calmness is movement. Calmness is freedom and freedom is

movement. Restriction of liberty disturbs and deprives man of calmness.

9.9138.

If you need to kill or create passion, imagine aversion or joy.

9.9151. Nanophilosophy is the philosophy of the north.

It's cold and difficult to live on the north. It's necessary to avoid mistakes in order to survive there. Nanophilosophy is the cure for stupidity and mental weakness, the elixir of common sense- a source of power.

9.9154. Synergy of thought.

The fusion of two thoughts generates the third integral thought, bigger than its sources and having new qualities and characteristics, that are different from the original ones.

9.9155. The river of thought.

By merging various sources of thought into one, you are creating a whole that is a lot larger than its details. The shores along the bed of the river of thought are fertile and rich.

9.9161. The chosen one.

The very fact that a man was born points to his chosenness. This sperm cell has defeated millions of his rivals and now has a great mission in life that he must accomplish.

9.9167.

All the unnecessary information is archived by the brain in the depths of the memory, so one should at first dearchive something unnecessary to remember it.

9.9175. A man is a small world.

Understanding of the world requires understanding of oneself. People do not understand each other because they do not even understand themselves. You cannot understand others without making sense of yourself.

9.9189. A woman's task is to create an atmosphere.

Women are similar to the Moon and the Moon is responsible for Earth's atmosphere. Consequently, woman are responsible for their family's atmosphere and viability.

9.9200. A beautiful language.

In English, many things are called by their names.

9.9204.

A new life usually does not start with a blank white sheet of paper, but with a blank black one, that is, from zero.

9.9217. Soft algorithms.

The patterns and stereotypes are the rules of behavior, programmed in the human conscience,. Robots can't break the programmed rules, but some people can.

9.9221.

Theory is strategy and practice is tactics. Tactics without strategy is dead.

9.9222. A man's strength of mind and spirit make him free.

A man is said to be what he is surrounded with. That's true, but a rational man always has the freedom of choice. A rational man can always do the opposite or, maintaining his neutrality, act absolutely freely.

9.9223. Philosophy is a life strategy.

Philosophy is a theory of life without which the practice of life is dead. Tactics without strategy is dead.

9.9224. Philosophy.

Life tactics are practical skills that are helpful in life and strategy is the foundation of tactics.

9.9237.

Beat foolishness with foolishness. Kill the dangerous and useless foolishness with the useful and safe one.

9.9242. Variable constant.

Human is basically a variable constant. Foolishness, laziness, fear, love, the other various emotions are always in his head and he can't get rid of them. But all these features of the soul can be either positive or negative. That is the same thing can be used both for the good and for the bad of the cause.

9.9252.

Usually everything makes sense - the lack of sense is a contingency.

9.9279.

It is very doubtful that time exists without movement. To want to achieve calmness is to want to stop time.

9.9283.

A new life usually starts with a blank black sheet of paper and has to with total destruction of old life.

9.9299.

Common sense is the comprehensiveness of thinking which does not exclude but includes all three extremes.

9.9329.

The speed with which the universe disintegrates is different in its different points. Consequently, the speed of time is different in the different parts of the universe.

9.9332.

Of course, wormy and rotten apples are whining noisily because that hurts.

9.9334.

An idea is all good but a deed is necessary to realize an idea. A man who managed to realize his idea is a hero who did a deed.

9.9343.

A man is not an apple – he may well roll over to another apple tree.

9.9346. Everything is just.

Some pretend to work and others pretend to pay them.

9.9347.

Injustice does not exist, but bad luck is common. When luck abandons a man for one reason or another, the poor fellow is in pretty rough shape.

9.9350.

Stores mainly sell poison and you buy and eat it because it's no big deal: people, just like cockroaches, are great at eating any poison.

9.9353.

Mirror neutrons deprive people of their free will.

9.9359. A strange construction.

If the speed with which the universe disintegrates increases, the speed of time increases too. However, if the speed with which the universe disintegrates exceeds the speed of light, the speed of time will start slowing down until time stops.

Evidently, the maximum speed in this universe is the speed at which time stops.

9.9360. The universe stuck in a time loop.

The excessive speed of light results in the time starting to go backwards. That is, if the speed with which the universe disintegrates exceeds the speed of light, time will go backwards and the universe will fold back into a point.

9.9380. You can eat anything, but don't overdo with it.

Eating little and often is useful in that all food is used up to refill the organism's current need in energy and fat – there is nothing to be saved for later.

9.9382. Obvious things are invisible.

9.9384. An electromagnet.

A man is, essentially, an electromagnetic system. Gravitation is a magnetic force that attracts all objects, and the moving man is an electromagnetic system that opposes the Earth's magnetic field. A small electromagnetic man is much stronger than the big Earth.

9.9385. Levitation prevents objects from merging.

Electromagnetic forces do not allow nuclei to get close to the point of strong nuclear forces starting to act. From this perspective, objects never touch each other and are always levitating.

9.9386.

Objects never touch each other. Firmness is the illusion of electromagnetic forces. You punch a wall and electromagnetic repulsion stops your hand.

9.9387. The soul is what gathers details into a whole.

The soul is an electromagnetic essence that ensures the integrality of man by gathering his atoms together. When his soul dies, electromagnetic power turns off and his body disintegrates...

9.9388.

In the Earth's atmosphere, a kilogram of fluff weighs less than a kilogram of nails. It's because the atmosphere is a liquid in which Archimedes' principle of upward buoyant force works. So, if the volume of fluff is bigger, the upward buoyant force will be stronger and pressure on the scales will decrease.

9.9390.

Apparently, gravitational and strong nuclear forces are the same thing, but at different levels.

9.9391. Pure intelligence.

Interestingly, man is an electromagnetic life form and computers are an electric one. Essentially, in creating electric life, man tries to create God...

9.9392.

A biofield is, essentially, an electromagnetic field that somehow transmits the encoded information, in the manner of technologies used in cellular communication or in wireless data transmission.

9.9394. An X man cannot be differentiated using Z.

Using a man for purposes other than intended is like multiplying by zero or differentiating E^x using Z.

9.9395.

A kilogram is the unit of mass and weight is measured in Newtons. This is why a kilogram of fluff weighs less than a kilogram of lead. Mass differs from weight on the Earth's surface.

9.9396.

Hypnosis does exist, as clearly evidenced by advertising.

9.9397. An exercise for weight loss.

Using brains intensively is more useful than running, since the brain consumes 30% of the organism's energy.

9.9398. Atmosphere is discharged water.

Lungs are modified gills and a man is a modified fish.

9.9403.

The body should be taken care of. The body is a very expensive and fragile machine that is hard to repair.

9.9409.

Everything that can be divided by three can also be divided by seven.

9.9410. The wings of love.

Social elevators are a myth that is not and has never been. To go to paradise, you should become an angel. Angels are the ones who have the wings of love. Love for truth, for your work and your dream.

9.9419. RNA organisms emerged from mud pools.

The Bible says God created man from the dust of the ground, that is, from mud. Scientists also believe that life originated in mud and clay pools...

9.9421.

An external source of energy is mandatory to defeat entropy, i.e. to put chaos in order.

9.9423. Evil Man.

A selfish person is an inward-looking system. Any closed-loop system is known to be inclined to the growth of entropy and transformation into chaos.

9.9431.

They like rice in Southeast Asia because of their climate. Torrential rains destroy all other plants, but rice grows waist-deep in water anyway.

9.9505. The biblical apple of sin.

Metaphorically speaking, all trees need sugar in order to incite animals to eat as many fruit and berries as possible and to spread as many seeds as possible while providing them with fertilizers.

9.9516.

The benefit – and harm – of everything on earth is greatly exaggerated.

9.9535.

The Earth may well be flat in a digital universe.

9.9612. Distance should be measured in hours, not kilometers.

Time should be measured in kilometers – time is speed.

9.9625.

The worst illness in paradise must be obesity.

9.9626. Particularly in paradise, everyone likes to eat.

Paradise is an illusion, otherwise everyone would die of obesity there.

9.9628. Love.

A woman is the source of energy for a man.

9.9653.

If you count on a contingency, God will stand in your way. If you count on a plan, your enemy is the Devil... However, your plan may have a place for contingencies, in which case both of these forces will be on your side.

9.9654.

Calmness is order. Any system is said to seek chaos, but any system seeks calmness too. The conclusion is: chaos and order are the same thing that exists simultaneously.

9.9655.

A system seeking calmness is simultaneously seeking chaos.

9.9659. Intelligence is the man's master.

The It is the man's master, but cunning and deceitful intelligence can easily deceive its master, if desired.

9.9687.

Get it right, it's like this: they catch, they milk and they eat.

9.9690. Meditation and trance.

Trance prevents you from going mad.

9.9692. In any good plan, there must be a victim.

In a good plan, there must always be something to sacrifice.

9.9693.

A victim is a must. No way without a victim.

9.9694.

A leader is not the one who pulls others after him, but the one who slows down and does not let others outrun him.

9.9703.

A person working for God is called an angel. Any person can become an Angel if he loves God. To love God is to serve God faithfully and loyally. Angels have very many responsibilities.

9.9704.

Angels are the first after God. Angels are God's will, God's word, God's servants, his hands, sword and shield. Angels are the strongest creatures on Earth.

9.9705.

Religions differ in the balance of rules and personal liberty.

9.9711.

Try not to harm good people, otherwise their angels will just grind you to dust.

9.9713.

People get fat from sins, that is why they cannot pass through the eye of a needle into paradise.

9.9714.

We move in time, not in space. Space is the derivative of time. In the absence of movement, both time and space collapse into a point.

9.9735.

Man often converses with God. Man's conversation with God is called a revelation. If you need God's advice, ask.

9.9740.

Reasoning is the process of man conversing with God. God converses with man through revelations.

9.9760.

Time is a very weird construction: it seems to be a fourth dimension, but it's not so... Our world is three-dimensional. Time is the derivate of a wholeness that was formed when the first three dimensions merged. Essentially, time is movement that is related to acceleration.

9.9762.

The world is falling into the abyss. That's true. Order always seeks chaos. On the other hand, while the world is falling into the abyss, the abyss is not standing still either because chaos is what seeks order.

9.9763.

The key point in hypnosis is not to stop: regularity turns speech into music, and the listener no longer sees into its meaning – he just enjoys the singing of birds.

9.9781.

Man is an electrical life form, so he cannot live without electricity.

9.9832.

God is law. God is the rules and laws by which our world is governed.

9.9833.

God is law. God is rules and laws that govern our world. The Devil is an exception to the rule.

9.9846. An absolute point.

An absolute point is an entity that has neither volume nor

area. The Universe originated from an absolute point and the absolute point is the smallest entity in this Universe. Any entity consists of absolute points. An absolute point can be folded in a point – an absolute point is an endless array of points. Any entity or even an entire universe can be folded in an absolute point. Any point can be placed within a point.

9.9847.

A man is the creator of the world and a woman is the creator of people.

9.9861. In transforming into an invariable, a man dies.

There should be an invariable and variables in a man. Variables are what allows him to grow and change.

9.9891. Sugar.

Fruit- and berry-eating by mammals somehow looks like cross-species sex. Helping plants to reproduce is such a pleasure.

9.9941.

The human spirit, soul and intelligence are an informational essence living outside man... These are books, ideas, reflections and words...

9.9952.

God is order and order is a plan. A man with no life plan is a godless man.

9.9963.

Paradise is possible only if there is hell. In the absence of hell, endless suffering will just drive people in paradise mad.

9.9970. God is intelligence.

Inside every man lives a devil, and only Intelligence enables the man to control him. However, as soon as the man loses his intelligence, the damned inside him wake up instantly and start doing evil. Mortal sins is what weakens or kills human intelligence.

9.9982. An increase in the speed slows time down.

Try to reflect as you go – that will save you plenty of time. On the other hand, the lack of time is greatly exaggerated, all the more so that time is elastic and can be stretched.

10.1332. Rational world.

Rationally, therefore, it is reasonable. This world is rational in all its aspects. Even accidents are rational, stupidity is rational, and love is rational.

10.1585.

Since time is an illusion, everything you do today extends to the future and the past at the same time.

10.1702.

The variability of things is related to the essence of their wave nature, cause-and-effect relationships in this matter are auxiliary.

10.1703. The triple system.

There is inevitably an intermediate link in the system of cause and effect. Whatever causes produce effects, step 3 is predetermined by the wave function.

10.1705.

The wave function is the solution to the egg and chicken paradox.
- What's the solution?
"The solution doesn't matter, just the frequency and the wavelength.

10.2024. The truth is the light.

The problem with the idealist is that he thinks the truth is absolute. In fact, truth is light, its nature is relative. Light is both a wave and a particle, dependent on speed and gravitational distortions.

10.2025.

Humility is a conviction in the wave nature of being. Love and joy cannot be killed, but suffering and fear cannot be killed.

10.2063.

The division of reality into positive and negative part there is a vivid example of duality, and of idealism.

10.2149. Wormy Apple.

Curiously, the inner voice you use to talk to yourself is not you, but the brain parasite that infected your distant ancestors, Adam and eve, when they ate the Apple the devil slipped them. On the one hand, the worm, of course, vile and generates a headache, and on the other hand, if not he was still sitting on a palm tree, like all the other monkeys.

10.2191.

The probabilities are not evenly distributed, but in waves.

10.2284.

Can we say that an ordinary two-eyed man is an idealist with a dualistic fragmentation of consciousness, and a one-eyed Cyclops is a monotheist, solid and solid as a stone?

10.2290. The saving fire.

The butterfly is a symbol of idealism. The flapping of a butterfly's wing can shake mountains. But nothing can save a butterfly from disappearing in an instant in the fire. Butterflies are dangerous, fire saves the world from the monstrous power of these monsters.

10.2327.

The essence of infinity is scalability. A dot, a ball, and the universe are all the same.

10.2328.

Infinity is limitation. A three-dimensional ball is limitless for a two-dimensional ant.

10.2337. Rehabilitation of idealism.

Idealists may be blamed for their attachment to the unimportant, but it is these small, inconsequential, but constant steps that ensure the progress of civilization.

10.2359.

Dualism, by dividing the whole into two parts, creates the illusion that they are two different parts. If you break these two parts into thousands more pieces, the illusion of difference will disappear. All the little pieces of right and left are absolutely interchangeable.

10.2389.

Dualism is a lie, because there is no difference between black and white, and anyone who walks the path from darkness to light will be able to see it for himself.

10.2551.

Our world is an electric car that rolls on the wheels of love, where love is a source of energy and Navigator in one person. It is desirable that the mind be friends with love. The elevation of reason breeds pride; the elevation of love transforms man into an animal. Need harmony.

10.2572.

If you want some new future, different from your present and past, you should forget and erase your past. The future is the past. To create a new future, the old past must be removed.

10.2587.

Good whether good? Is the child of evil good? How long has he been kind? Will it necessarily turn into evil?
- The point is, the children of evil are the wheat and the chaff. The tares will become evil, and the grains are what our world exists for.

10.2673.

The unity and struggle of opposites is not the struggle of evil and good, but the struggle of opposite types of evil. Evil hates each other, but being unable to separate itself from itself, lives and suffers.

10.2686.

There are no old or new truths, no time at all. Time is a delusion of the mind. You can be aware of the truth, you can be unaware of it. The truth has existed for billions of years, and you realized it today. Is this a new truth?

10.2910.

Zero is something that cannot exist without external energy. A unit can exist and grow without external sources of energy.

10.2911.

A lie is something that needs external energy to exist. The truth is that external energy is not needed and can even be a source of energy itself.

10.3060.

Wolves are useful to sheep because they protect against other predators and regulate the population, preventing the exhaustion of the food supply and the destruction of the ecological niche in which sheep live. The wolves protect the sheep from themselves.

10.3247.

The truth is the reality. A lie based on this reality plays games with another lie. The truth is the basis of a lie the add - in is addicted to it. Truth is the ocean, falsehood is the air.

10.3316.

Truth and falsehood are light. You remember that light is both a wave and a particle.

10.3408.

The uncertainty of being has a shifted center of gravity at the level of the Golden section. The randomness generator gives 62 negative outcomes, 37 positive, but in the long run this probability gives a benefit of 300% for those who serve love and truth.

10.3439. Spiral of truth.

The number of variants of truth is infinite, but the number of basic elements of truth is limited. For example, the periodic table contains 130 elements. Perhaps there will be new elements in the next round of existence. And so on.

10.3495.

Since God is everything, then God is both a point and a universe. You can take any point and it will be God. If you don't want to take the dot, you can take whatever you like.

10.3501.

God is an absolutely scalable entity-a point and a universe at the same time. This is the nature of truth, love, and order. In fact, you can take any thing and love to scale it from a point in the universe.

10.3548. The theorem of the point.

Although various sources claim that the truth is incomprehensible, I do not agree with this. God is everything. God is the truth. All this means simultaneity and scalability. God is both a point and a universe at the same time. Therefore, by understanding the point, you can also understand the universe. At the time, Mr. Malevich drew a Black square, but to draw it, he needed 7 volumes of the theory of Suprematism. I want to draw a point and for this purpose I am already finishing the 10th volume of "Variothoughts". The main point of "Variothoughts" is to solve the point theorem by proving that a point and the universe are the same.

10.3550.

How do you separate creation from destruction? The artist takes a pencil and destroys the perfection of a clean white sheet. "Scoundrel! - in hysterics shouts idealist. "That's sacrilege!"»

10.3552.

The universe is a black dot on a black background, and there is no light. Light is the illusion of the mind.

10.3565.

What is destruction and creation? Dream after dream?

10.3676.

Uncertainty and randomness is a mistake. However, any living system must work with errors, because error is the basis of evolution, growth and life in General.

10.3677.

God is perfect, but mistakes and uncertainty are part of perfection. The system is designed so that errors do not kill it, but generate novelty.

10.3682.

Truth is a solid state of energy. Matter in the universe is only 1%, but this does not mean that the rest of the universe is not.

10.3730.

Fish live in dream under water. The real above-ground world hurts Pisces. As the glass fish walking on land, the pain makes them Wake up.

10.3998.

The brain is an electrochemical system, a biochemical reactor, essentially a living microprocessor that never shuts down.

10.4153.

The world is intelligent, the only one who is not intelligent in this world is man. The meaning of human life is to strive for intelligence.

10.4291. The system of chaos.

Chaos is a unity in the same space of time from many of the ordered processes.

10.4359.

Humility is what gives form to things, that is, turns them into truth and beauty.

10.4435.

Truth is a systemic process and order of things. Love is the force of motion that makes things move according to plan. Truth is when everything goes according to plan. Truth is a process. Truth is when things are in their time in their place.

10.4436. The idea of things.

The form of things is only the idea of things. Form defines its content because it shows them a purpose for self-organization. The elementary perfect bricks of being, having received the purpose, begin to be built in its form.

10.4437. Nice uniform.

A being is something that has an idea of its form. And if this idea is sufficiently perfect, limited and similar to truth and beauty, it has every chance to attract the content.

10.4557. Three extremes.

They say extremes are evil, but they don't say this figure is a triangle. And what is called the middle is only the third extreme.

10.4576.

From the point of view of quantum physics, time does not exist, and there is no problem of eggs and chickens. And since God is everything, there is no problem of the first cause. Billions of years will pass and people, as a society, will become one whole God. We seek unity with God and with each other. Having achieved perfection and unity, God will create the universe anew, restarting the process in an endless circle. The point of re - creating the universe is that perfection is death, so when perfection is achieved, everything has to be erased.

10.4636. Computer simulator.

Inside, the brain creates a toy universe where the toy Self

plays with toy copies of people from the real world. The subconscious, watching this game, makes some conclusions on the basis of which behaves in the real world.

10.4637.

Do you know where idiotic unrealistic expectations come from? Your toy Me, playing with the simulator of the real world, makes some stupid predictions, and you, like a fool, then run and wait for them.

10.4647.

They say that the problem is easier to prevent than to solve... Lying.
Problems are very conditionally connected with the reasons. If you kill one cause, the problem will find a dozen new ones.

10.4678.

Consequences are not related to causes. Moreover, there are no causes and consequences. All entities are self-sufficient and simply motionless next to each other. Entities choose their own causes and consequences. The consequences of this goal. Choosing a goal, the entity, paying attention to any entity that is near, makes it the cause.

10.4679.

What seems chaos from below is absolute order and clear regularity from above.

10.4680. Liars suffer greatly.

Deceive nature will not work. By deceiving nature, man punishes himself. Barriers lie like steps, at each level, the punishment becomes worse. A man might die quickly of some common disease, but by defeating them he would meet with cancer. Cancer causes him to die painfully. After defeating cancer, a person will meet something else that will make his hair stand on end with horror. And so on. And so everywhere. You can cheat the system, but it is unbearably painful in the consequences.

10.4718. the Main question of Syntalism

Truth does not need to be sought or proved. Everything is the truth. Take whatever you want, it will be the truth. Of interest is the proof that two entities that seem incompatible and different are actually exactly the same.

10.4736. Energy, truth and lies.

The meaning of Yin and Yang, truth and lies is that we live in a pseudo-three-dimensional world. There are three truths and one lie in this world. Time is a lie. Time does not exist. Time is the process of moving the three truths. But even three-dimensional truth is only information about the form of energy.

10.4740. The algorithm of truth.

If the Trinity is viewed as a transistor, then God, who is the universe, is the unity of trillions of transistors. But being just a microprocessor is boring, you need some purpose. God, to not to be bored, created itself algorithm life. Now God has a job, he is busy endlessly calculating the algorithms of life and generating more and more new forms of being.

10.4750. Unemployment in Paradise.

The picture of existence about the following. There is hell, this is earth. There is Paradise, this death. The soul, like a wave of light, half the time living in hell and half in heaven. The process of rebirth goes on endlessly. The caveat is that if you exhaust all your capacity for joy and pleasure in hell, the time spent in Paradise, when bliss is no longer happy, you seem the embodiment of a nightmare. Therefore in hell it is better to show moderation to pleasures. Moreover, hell is a kind of resort, where the gods are granted such happiness as the opportunity to work. Work is the highest value of hell. Mired in the endless boredom of the Gods the soul is ready to sell to the Devil for the right to work.

10.4865. Super LEGO.

Since time and space are illusions, any imperfections are just pieces of the puzzle scattered in space and time. The meaning of the game, collecting puzzles into a whole, from the details to collect perfection.

10.4894. Will you be my reason?

In fact, cause and effect are one and the same. Things desire to be and therefore create their own causes. It's like casting. The investigation looks to the past and pokes a finger at everyone it likes.

10.4897.

The form determines the content. But form and content are one and the same. Active stance this when you-form of, passive stance - when you content.

10.4928.

The man is quite wonderful comes up with his love, and then implements it in reality, so turning the lie into reality. Moreover, the lie does not exist, because the lie is also a reality, only small, like a current pulse running through the neural network of the human brain.

10.4946.

Order is a machine that produces chaos out of chaos. And it is an amazing and magical machine that produces more than it receives. A machine that produces thousands of times more nothing out of nothing than it was.

- How can you magnify anything?
"Well, it's all an illusion.

10.5033. Super team.

When you download a problem into your brain and start thinking hard, it's like billions of neurons in your brain start calling to each other, asking... What to do? Any ideas? What are the options? And billions of voices whispering and arguing... until they find the perfect solution.

10.5039. Good and evil.

Idols and idolaters is a closed ecological system of trees and animals (insects, birds, animals). Trees dream to grow, to reach perfection, to begin to bear fruit and to become idols. Animals dream of eating and having fun. Animals eat plants, spread their seeds and die. The roots of trees devour dead animals. The forest of a thousand idols, some say it is God, others say God is everything.

10.5054.

Varieties of is full of idols, idolaters, too, are different. There are three main types of idolaters: animals, birds and insects. Idols are fruit and just beautiful. Predators are especially useful to idols as sacrifice to them, for this idols are feeding these victims.

10.5064.

Suffering was created so that man could love the real world. It is the love of life that saves suffering. What is the love of life? It is movement, creation and cognition.

10.5132. Meaning of life.

This world was created for human happiness. Although life is suffering, this suffering is needed to motivate you to find love. When a person finds love and begins to serve it, his life becomes happy. Love is man's mission in life.

10.5137.

Nothing contradicts anything, because life is a paradox and a unity of opposites. One doesn't interfere with the other.

10.5152. Information about the form of energy.

If time is movement, and technically, time does not exist, then is there a motion? If we look at movement, there is no time. If we look at time, the movement is the passage of time, then there is no space. Time and space are one and the same entity, which differ only in the method of measurement.

10.5223.

If you limit your fantasies, you will turn them into reality. Reality is an illusion limited by form.

10.5224. Nothing and Nothing.

There is only energy and information. Information is an illusion. But I'm telling you, there's nothing but Nothing and information. Energy is also an illusion.

10.5314. Willpower is strength of mind.

In addition to the point of view and infinity of chaos, there is a third entity, which is limited chaos. Limited chaos is the truth. Truth can be like a circle, a square, a triangle, or any other geometric shape.

10.5514.

The real world is energy limited by form. Form is an entity limited by time.

10.5515.

Energy, form, time. Form enslaves energy, and time frees energy. Form is space limited by time.

10.5528.

You can't cheat the system, because the system is the truth, and a lie is something that doesn't exist.

10.5531.

Time does not simply liberate energy from the bondage of form. Time is the energy that gets tired of being in the same form all the time.

10.5547.

On paper, everything is smooth, because paper is two-dimensional, and real life is dynamically three-dimensional.

10.5564.

Life is a system with a shifted balance towards the Golden ratio.

10.5581.

The interpretation of Truth and the nature of Being given by Syntalism is objective and is in good agreement with all external sources and reality. We don't see it as vulnerable or illogical. If you find it, fine, we'll close the gap with another logical construction. The entire Syntalism knowledge system is well-founded, connected to each other and thousands of external sources of information.

10.5622.

In the human eye, sticks are responsible for distinguishing black and white colors, and three-dimensional cones are responsible for color. To see the color world, you need three-dimensional thinking.

10.5625.

The same thing happens to the brain of an alcoholic that happens to yeast in the preparation of wine. First, the tremors devour all the sugar and rejoice, producing alcohol. Alcohol is a fine solvent, which then dissolves, like sulfuric acid, the one who gave birth to it.

10.5633.

At the micro level, perfect order reigns. At the macro level, the options of being and free will becomes much more.

10.5681.

Immobility in real life is impossible, for life is movement. But in our lives possible a squirrel in a wheel. A person is fixated and runs around the circle of his sins, this state can be called static.

10.5720. Dwarf idol.

Light is a lie because it is an illusion of the mind. The mind, using such a sense organ as the eyes, received some information and decided that it was light. But the eye is limited. The light limited by the eye is just a tiny tiny bit of real light, only 1% of The sun's radiation.

10.5737.

Life is suffering, so that people do business and not sleep. Pain awakens the mind. But in order not to go crazy from the pain, we need love and humility.

10.5777.

People are like atoms. The forces of mutual attraction and repulsion form a society out of them. Society is a chemical element of the periodic table.

10.5795. No random mistake.

The human mind is the product of a system error. Man

should not have existed. Man is the only element of imperfection in our perfect world.
"Was the mistake accidental?"

10.5802.

The point is that the cause produces the effect, which, growing up, becomes the cause of its cause. This is the unity of Yin and Yang, good and evil, darkness and light... and even our universe was created by this function.

10.5803. The wave nature of cause and effect.

When two entities are close, they are like wave objects at a constant frequency begin to change, one, turning into a cause, the other-in effect, then Vice versa.

10.5841. Movement is life.

The only option is the generation of carrot and stick. Fear and desire to run away from him. However, there is a third option, it's pretty wacky, and it doesn't seem to solve anything, but it creates movement. The movement of the three colors is a basic color picture of being called life.

10.5849. Four Islands and one Peninsula.

The periodic table lacks three Islands of stability of chemical elements. The first island in district 130, then on level 140 Yama. The second island is level 160 and around it. At the level of 180-200 pit. At level 200-220 there is another, the smallest island. The Peninsula is level 110-120 (radioactive isotopes).

10.5946. A perfect world.

Studying the biology of living creatures, I was struck by the ideality of their device. They are monstrously complex and perfect. So I wrote the word and you read it and understand something. For that to happen, billions of chemical reactions have to happen. Can you imagine the billions! And there is not a single failure! And it works everywhere! In this world everything is infinitely perfect.

10.5948. The Trinity is the point form and the universe.

The forest is a democracy of dots. All points are equal, and the forest is unity. However, in the system of these two there is also the concept of form, that is, certain associations of points on formal grounds, that is, groves. Our material world is only millions of different forms, which together are the forest. The form is a virtual Union of points. Form is illusion, all forms of the material world are illusory.

10.5949.

From the point of view of the point and the universe, the forms are equal. One form is no different from any other form. Form is life. This is an endless movie, some scenes of which do not have any special meaning. Points are equal between themselves, all point this universe.

10.5950. Evolution of the universe.

God is the standard that has reached perfection. That means God stopped evolving and died. God is dead. It's sad, but ... he didn't die in the past, he died in the future. And this, of course, pleases. In the present, God is still imperfect and continues to create himself with his own hands. The perfect God, who died in the distant future, creates himself from the future of the past.

10.5951.

God is the standard of perfection in the sense that the basic laws of existence are unchangeable and do not evolve. Only forms evolve. Only the image on the TV screen changes, the TV itself does not change. In other words, God is asleep and the world is dreaming. When God wakes up, the dream will end.

10.5952. Endless fantasy.

The forest itself is a tree in relation to those above it. In relation to the earth, the forest is equal to the ocean and the plain, mountains, winds and clouds, etc.and the earth is equal to other planets. And the sun - other stars, and galaxies-galaxies.... Who is the universe equal to?
"Another universe, apparently.

What makes you think man is divine? Maybe it was invented by the planet Earth, and the earth was invented by the Sun, and the Sun by the galaxy. A man descends still lower and comes up with virtual fantasies and dreams. Creates computers and virtual games to escape into. And within these game-worlds will emerge new worlds, where such same, as and you, start create their even more deep worlds.

10.5960.

The world is a color movie, not a black off TV screen. It is enough just to know that the TV can be turned off and on. You can switch channels or even shoot your own movie. And by the way, turning off the TV is not a problem. Both turned on and off. The point is not to achieve dissolution, but to know that a circle and a square are one and the same. What a wonderful prove, for example, mathematics and the equation of squaring the circle.

10.5962.

The universal method of knowledge is called love. What is love? Love is attention. Pay attention to something, start thinking about it, and, sooner or later, get an intuitive answer. An important moment, intuition arises not miracle. Intuition is the result of learning the neural network of the brain. It is possible to train a brain theoretically and practically. There is mathematics and there is physics, these two complement, but not deny each other. Practice is personal experience. The theory of reading and observation.

10.5963. Divine reality.

This is the real world and our universe, in which the laws of physics and Cybernetics work. All sorts of fantasies that people come up with are virtual worlds generated by the brain within itself. In the real world on water cannot be walk, and in dream can be. Mr. Jesus, when he walked on the water, wanted to say, gentlemen, Wake up, you're asleep.

10.5970.

The truth is the information that gives form to energy. In fact, it is a form of self-organization of energy. There is only energy and its illusions. Energy likes to be beautiful, so truth is beautiful. Ugly truth is difficult to get energy. Energy loves beauty. Love is what connects energy and beauty.

10.5978.

The forest is the trees, the unity of the point and the universe. Different combinations of points give rise to different forms of the real world.

10.5979. Illusion of forms.

Kill the lie and illusion that any material form is an illusion, so life is so afraid of time.

10.5985.

The more persistently a person tries to defeat time, the faster time will destroy him.

10.6009.

I have noticed that evolution does not always follow the most obvious path. The usual path of evolution is a strange and difficult path. The one who is difficult evolves, not the one who is easy.

10.6068. Clam.

It's not the easy ones who win. Those who find it difficult win, but they don't give up. Those who are easy, relax and turn into clams in the shell.

10.6069.

The evolution of annelids into molluscs is a creative dead end, telling us the fate of those who are easy. It was harder for ordinary worms to live, so their evolution was more varied and significant.

10.6108.

If man, as a biological species, did not rise up and kill all the competitors, it might give the monkey a chance to become a decent person, go to College ...and all that. And so... well, bad luck to the monkey, what can you do?.. Even 50 thousand years ago, there were 7 types of people ...now there's only one left... And how many of them before that was unknown at all. Homosapiens believe themselves to be gods and kill and eat everyone else ...and in General... Homosapiens is the most vicious and monstrous predator in the history of the planet Earth. I think the monkey needs sympathy, it just had bad luck with the neighbors.

10.6129.

The evolutionary process didn't care about the future. It works here and now. The main thing is that now. Let the day of tomorrow take care of itself, we must take care of the day of today.

10.6179.

The point is also too much, only in the opposite direction. If a point is turned inside out, it becomes the universe.

10.6183.

Baboons on some spiral of evolution far outstripped man, which killed them. Man with difficulty, but continued to evolve, and the happy baboons went the easy way.

10.6185. Love is life.

Love is an ecological niche. A person falls in love and this becomes his source of energy. The loss of love creates depression, a person metaphorically dies, but then finds a new ecological niche, and resurrects body and soul.

10.6189.

They say that man (and all other animals) is just a protein shell for the ancient anaerobic microflora that lives in his large and small intestines. In fact, anaerobic bacteria have specially created a protein shell for themselves, so that it gets food and protects it from oxygen. When food enters the stomach, the bacteria process it and give some of the energy to their protein symbiont.

10.6190. The Queen and king.

Man is a living anthill, a symbiotic colony of anaerobic bacteria that have created a living environment that protects them from the poisonous oxygen atmosphere. The subconscious is the Queen of the hive, which controls everything through the nervous and hormonal system, and the consciousness is an additional analytical module-a kind of state Duma.

10.6202.

Natural selection in human society works cunningly. The tares kill themselves, and the grains grow and thrive in the absence of strong competition.

10.6249.

There is nothing wrong with so little truth and light. 1% of light in the realm of 99% of darkness is the normal state of things. You don't need a better one. The best is overkill and the enemy of the good.

10.6289.

One consists of zeros. A unit is a column of parallel rings-zeros. One is a zero that manages to either grow in a spiral, or climb from ring to ring. Each ring is an energy level that advises the level of energy available to the system.

10.6297.

Zero is a flight in orbit. Different orbits correspond to different levels of energy supply. The complexity of orbiting is the same everywhere. The transition from one energy level to another requires effort. Usually, the transition of an electron from orbit to orbit is associated with its self-destruction and re-creation.

10.6318. Error-resistant system.

Nature is an ideal system in which, on the one hand, there is nothing superfluous, and on the other hand, there is a special redundancy and uncertainty. This is done in order to make the system scalable and resistant to errors and destruction of critical parts of it.

10.6348. Solid-liquid essence.

The harmony of being is at the level of the Golden section, not the middle. A color world could be constructed with three colors, but reality uses four basic chemical elements. The fourth element makes the system invulnerable to errors. Pseudo hardness. A pseudo fluid. Usually the system is solid, but when the hardness is destroyed, the system becomes liquid and adapts to the destruction, restoring integrity.

10.6580. There is no death.

Society is a mycelium, and the human individual is a mushroom. The death of an individual mushroom means absolutely nothing, because all the information is stored in the mycelium and the correct answer to the question "who Am I?"it sounds like," I'm a mushroom, but I'm also a mycelium.".. It is extremely difficult to destroy mycelium, they live for thousands of years.

10.6582.

What makes a person related to the idea of God is that the human brain creates the universe within itself and all its inhabitants, including itself. The brain is a kind of God in miniature, a child of God. And the human mind, unlike a squirrel, can not just reflect reality, but create its own images and then transfer them to the real world.

10.6583. Humility and pride.

The difference between a human society and a mycelium, hive, or anthill is the comparative isolation and independence of individual individuals. The hive is egoistic and full of pride, which creates in it fear and isolation from the world. People are more open and this gives them certain perspectives.

10.6584.

Animal this is a hive, not a bee, a mycelium, not a mushroom. So in the human community, the subject is the society, and the people it's just mushrooms and bees.

10.6594. Eat me.

By the way, the more beautiful the mushroom, the faster it will be eaten by animals, and by mixing spores with shit, they will contribute to the reproduction of mycelium.

10.6596.

Potential is no better or worse than reality, it is just another form of reality. The reality tends to accumulation and the evolution of forms, nothing more.

10.6597. Theory and practice.

Children play their computer games, and father God looks at them and smiles. Let them train, the game will prepare them for real life. The main thing is not to get stuck in your games for longer.

10.6613. The All-Knowing God.

God's omniscience is the concept of mycelia... Every person is a mushroom, but every mushroom is a mycelium. All that each individual mushroom knows, the mycelium also knows. Mushrooms, ants, bees and other animals are not independent living organisms, but parts of larger living creatures. At a level even higher, the societies themselves are United in a kind of super-Society that unites all forms of life on earth. You can go even higher.

10.6615.

Evolution constantly comes to a standstill, but does not give up...

10.6624.

Human society is something like a form of evolutionary development of mycelium.

10.6637. The dungeon of the mind.

Ideas don't come from us, but they don't come from the sky either. Man is a fungus, part of the mycelium.

10.6657.

Dolmens on the black sea coast are a symbol of the mushroom and lingam. The fungus is the sexual organ of the mycelium. Metaphorically, a dolmen is a penis of God, a symbol of fruitfulness. The hole in the middle is a hole for sperm.

10.6661. What are you doing for the system?

The life of a single-celled creature is absolutely unbearable. Single-celled ones are the most miserable, they are eaten by everyone, and they are no one. The life of a small creature is very dangerous, as Mr. Piglet liked to say. Getting big and strong is a great idea. Multicellularity allows you to eat everyone and not be eaten yourself. All life is a complete symbiosis. Man is a complex symbiotic Union of a huge number of living organisms, bacteria, amoebas, fungi and viruses.

A slime mold is a situation where a bunch of different amoebas join together and form almost a single organism in order to start bearing fruit. Individually, amoebas can't even reproduce. But the slime mold is not a mushroom, it has no real fruitfulness. The slime mold doesn't have an internal rod, and that's a problem. We need a skeleton, or at least a mycelium.

The philosophy of Syntalism is a beautiful skeleton around which you can build a large and strong symbiotic organism. Symbiosis is a synonym for the words "life" and "synergy". We need symbiosis for survival and reproduction.

10.6678. The law is common to all.

Nature is very persistent. Evolution has tried to create a living cell many times, and it only succeeded on the 6th or 7th attempt. All previous attempts ended in extinction. The point is that if you don't pass step One, you can't go to step Two.

10.6679.

Mushrooms are different. Yeast is a single-celled fungus. Mold is an imperfect fungus. Noble mushrooms are porcini mushrooms, morels, chanterelles, etc.

10.6680. A successful symbiote.

Lichen is one of the most successful slave owners in the world. He has enslaved a lot of cyanobacteria or algae, and forces them to release energy from the sun while he lies in the sun for thousands of years. Lichens live for thousands of years and do not know grief.

10.6683.

The mechanics of evolution is the process of exchanging information. The exchange of information takes place in the process of creating symbiotic unions in order to more effectively survive, defend and extract energy in a changing competitive and aggressive environment.

10.6685.

The Internet is somewhat similar to mycelium, it permeates everything and unites mushrooms. The system is full to the ears of all sorts of fantasies, one bad thing-passive and motionless.

10.6687. The Trinity.

The ocean is a symbiotic organism, the unity of all living organisms that fill it. However, this is not all. The atmosphere is the same ocean. But this is not all. Underwater, above-water and underground combine into one single symbiotic organism.

10.6688.

The complex consists of the simple. DNA of the fungus and

human DNA is similar to 55%. This means that, having created the mushroom, evolution has already passed more than half of the way from Nothing to man. For example, a RAM is already almost 90% human.

10.6690.

The weak evolve through symbiosis, which is metaphorically love and synergy. The strong evolve at the expense of parasites and viruses. Strong in the process of fighting viruses, fungi and parasites integrates their DNA into its DNA. In human DNA, for example, about 30% of the DNA of viruses.

10.6691.

The strong weakly strives for symbiosis, because symbiosis is a Union of the weak in order to survive and become stronger. Strong so strong, they have no special reason to unite.

10.6693.

The weak are more capable of evolutionary growth than the strong. The weak are able to unite, consolidating their best evolutionary advantages and findings.

10.6699.

The worm Apple is a symbol of a fly in the ointment. Curiously, from the point of view of nature, a worm-eaten Apple is a great idea. In fact, it is a fruit with animal protein, a priceless treat for all animals.

10.6700.

Wormy apples are more valuable to nature than non-wormy ones. Worm apples contain animal protein, which makes them particularly attractive to all types of animals.

10.6701. Meat apples.

Worms and wood are an example of a tricky symbiotic Union. The tree is very profitable to produce meat apples.

10.6705.

The earliest inhabitants of this planet are anaerobic bacteria. Second came their deadly enemies, the mushrooms. The third were viruses. However, it is possible that the viruses were the first, because they are not living viruses. A virus is a transitional form of life from the inanimate to the living. Human DNA is 55% similar to fungi, 30% to viruses, and the rest is probably due to bacteria.

10.6706.

In fact, the evolution of life took place in three stages. First there were viruses, then anaerobic bacteria, then fungi. The mushroom is very similar to the crown of evolution, because all the other stages of life development are already a superstructure over this Trinity.

10.6707.

Fungi parasitize bacteria, grow them, and feed on them. Viruses are parasitic and live inside bacteria. On the one hand, bacteria are victims, and on the other hand, fungi feed them, grow them, and organize them. And the viruses that live in them, in General, became the primary cause of their appearance and evolutionary development.

10.6709. Evil under the mask of good.

Infected bacteria produce signal interferons to let other bacteria know that they are infected with the virus. However, many clever viruses have learned to block the production of interferon, which allows them to infect neighboring bacteria without provoking the immune system to resist.

10.6713.

The Internet is very similar to mycelium, and the people sitting in it are like mushrooms. But mushrooms are different. Ants, bees etc also mushrooms.

10.6714.

This world is ruled by mycelial fungi. Mycelium is very tricky. Mycelia grow and feed trees and plants to then force ants to feed and grow mushrooms.

Everyone eats mushrooms. They fatten animals with plant food and eat meat. They force animals to eat plants, and inside them they eat them. Ants feed mushrooms with leaves. Termites-cellulose. Animals feed mushrooms with grass, fruit, plants, and meat. Mushrooms, penetrating into the animal, determine its behavior. Determine what to eat or who to be eaten by. The only enemies of fungi are bacteria and viruses.

10.6715.

Curiously, without mushrooms, plants and animals will not be able to eat and live. In fact, mushrooms grow plants and animals to eat.

10.6716.

An amazing fact, but plants do not know how to feed themselves, they are mainly fed by mushrooms. The fungi themselves feed on dead animals and plants. In fact, fungi and bacteria specifically created animals and plants for food and protection from oxygen. The viruses in this system are very strange creatures that were probably the very first to create bacteria. A virus is a pseudo-living form of life that cannot exist outside of bacteria. Viruses are not alive at all. Viruses are like the very first, transitional biological form from inanimate to living.

10.6717. The Holy Trinity.

The battleground of planet Earth is a war between fungi, bacteria and viruses.

10.6719.

The Internet is passive and resembles mycelium. Mycelium turns people into mushrooms, and passivity fills them with fear, lies, and fantasies.

10.6752. Free people.

In a three-class society, there are lower bacteria and single-celled ones, which are dominated by a complex hierarchy of fungi, headed by a higher mushroom. The third force is viruses, those who did not want to live in the system of slaves and masters, but they can not live without them. Viruses are nihilists and free people who don't like this whole system of slavery and want to destroy everything.

10.6755.

Life is a symbiosis. There is no life without symbiosis. To survive, we must form a single symbiotic organism. The life of a single-celled creature is unbearable. Anyone can eat it. We must create a multicellular symbiont society.

10.6763. Don't deviate from your goal.

Observing the evolutionary processes, I noticed that nature never gives up. The idea to create mushrooms was attempted several times, and ended in extinction. Only 3-4 attempts were successful.

10.6766. The problem of choice.

In essence, all human actions are perfect, and it is impossible to do better than he does. Our world is so cunningly arranged that no matter what a person does, it will be, in any case, a great solution.

10.6768.

The causes are not important, the consequences are not important. Nature and evolution are interested in processes occurring here and now. It is important to survive, reproduce and create now. What happened yesterday and what will be there tomorrow-we are not interested in much. The past has already taken care of itself, and the future is well spoken of in the gospel.

"So do not worry about tomorrow, for tomorrow will take care of its own: enough for each day of its own care."

10.6780.

It is not the individual that evolves, but its DNA. DNA is the host of the body. DNA evolves and changes the body for itself. First you change yourself from the inside, and then everything changes from the outside.

10.6783. Love is sharpened attention.

The human mind was created by mushrooms. Mushrooms allow, endlessly focusing, to see the nature of things, the nature of fire, water, wind, sky ...the universe.

10.6878. The virus of evolution.

It is not a person who is evolving, but viruses and problems that want to devour him. Fighting viruses, a person learns about them, incorporates their DNA into their DNA, and defeats viruses with their own weapons. Altered human DNA changes human nature. Viruses fall back and come up with a new weapon, attack again. Thus, the more problems a person solves, the more he mutates and evolves.

10.6879.

The reason for evolution is viruses, which, in fact, are elements of inanimate nature. Thus, the factor of inanimate (pseudo-living nature) is the main engine of life. In fact, inanimate nature is evolving in order to destroy the living.

10.6888.

Pride is contempt for the inferiors. The devil wants to be the best and despises everything small. Real angels are tiny accidents, specks of dust and atoms. God is at the bottom, not at the top. The universe grows from a point, not degrades from somewhere above.

10.6911. Black square.

As Syntalist, I do not believe in mysticism, I believe in quantum physics, chemistry and cybernetics. For me, someone else's soul is not dark, but a black box, a cybernetic artificial intelligence system. However, I see the universe and its God in the same way.

10.6917.

According to Syntalism, man was banished from Paradise for humane reasons, because Paradise and reason are incompatible. In Paradise, the mind degrades and dies. The mind is that which overcomes problems and obstacles. In conditions of complete relaxation, higher fungi degrade and turn into single-celled fungi, a typical example of this is yeast.

10.6956.

The shape of the glass has an effect on the taste of the liquid in it.

10.6980. Love is Aikido.

Human DNA is 30% made up of the DNA of the viruses that attacked it. Getting to know your enemies, a person uses their weapons against them.

10.6998. The yeast fungus.

Murderers evolve, the evolution of victims is secondary and based on the knowledge of their enemies. If you destroy the killers, enemies, and problems, evolution will end, and the victims will degrade into single-celled yeast.

10.6999.

Evolution is not a constant, but a dynamic process. Species either progress under the influence of problems, or degrade under conditions where everything is good and fine.

10.7000.

Paradise consists of the simplest single-celled creatures, so when you have no problems, you start to endlessly degrade.

10.7001.

The devil is all so complex and perfect from the fact that he lives in hell, where someone constantly wants to eat him.

10.7002.

Since the main engine of evolution is inanimate nature (viruses and circumstances), we can say that the cause of evolution is accidents, errors and space-time limitations of existence.

10.7005.

Civilizations are dying out for two reasons. Or they will degrade into a single-celled Paradise. Either technology and complexity reach the point where hell becomes unbearable. The unbearability of life in hell leads to the self-destruction of the system.

10.7031.

The victims are encouraged by viruses to improve their weapons. The more appetising the victim, the more you want to eat it.

10.7032. God's sheep.

The symbiosis of sheep and wolves is that the stronger the wolves become, the more appetising and numerous the sheep become. Viruses train and contribute to the evolution of their victims. The more effectively wolves attack sheep, the more effectively sheep evolve.

10.7035.

Life in Paradise inevitably dies, because shivers, eating sugar, produce alcohol, which then kills them.

10.7071. An exotic monkey.

It is a lie that man is descended from an APE. Man is an APE, just one of hundreds of breeds of apes.

10.7073.

Curiously, viruses and diseases are more to blame for human evolution. Having mastered the fire, the person became quite tender and painful. More diseases-stronger evolutionary progress.

10.7074.

Black from white is about as different as a monkey from a human. That is, it seems different, but the evolution was so gradual and imperceptible that you can not say that once a mother APE gave birth to a son of a man.

10.7089. Cream and tomato sauce.

We have already found that fear and love are basically the same thing. The only difference between them is the emotions they generate.

10.7110.

The present doesn't exist. What you think is the present is

the past. You live in the past and dream of the future. Space is three-dimensional, and time is two-dimensional. The paradox is that the vulgar is no longer there, and the future is not yet there. In fact, you live inside your illusion.

10.7114.

Since time and space are illusions, God, who is a point, can afford to be everywhere and always at the same time.

10.7129. Three in one.

God is both big and very small. God is a huge predatory mycelium, the owner of the forest. God is a harmless little mushroom. God is all the inhabitants of the forest who feed at his expense and serve him.

10.7133. Citadel of darkness.

Mushroom mycelium is very similar to the brain, it also lives in the citadel of eternal darkness. Mushrooms are the nerves. Trees, grasses, insects, and animals are its symbiotic bacteria and sense organs.

10.7134. The dead souls.

When people die, their souls become part of the mycelium. Everything that rots and is eaten by mushrooms transmits its information to the mycelium. Over thousands of years, mushrooms have accumulated great wisdom. The souls of the dead live in virtual universes. Trees are solar-powered servers that are tended by bots, insects, and animals.

10.7137.

When a living thing dies, putrid fungi come and read all the information from it. In the future, all this information goes to the mycelium, where the mycelium restores its information model inside its information system. Mycelium is like a brain, it builds a model of the real world inside itself, plus it invents thousands of different worlds of its own.

10.7139.

Dead souls don't live in mycelium forever. Mycelium controls not only the dead, but also the living. All living things are in a symbiont with mushrooms. For example, placental procreation in animals is the result of symbiosis with fungi. In other words, fungi can overwrite embryos by setting the properties they need.

10.7146.

Women are slightly closer to mushrooms than men. The female placenta and the ability to procreate are the result of symbiosis with fungi. That is why a woman likes to plant and grow something in the garden. A woman is like ants that feed and grow mushrooms. All plants are symbiotic organisms of fungi.

10.7147.

Mycelium programs human consciousness at the embryonic stage, during the process of fetal maturation in the female placenta. The placenta is, in fact, a part of the mycelium, inside which it piecemeal collects a new person.

10.7149. Immune protection.

It is best to protect yourself from the enemy with their own weapons. Than it attacks you, pull it out, and shout: "This is mine!"Viruses are powerless against their weapons.

10.8411.

Light is not a two-phase wave system, but a three-phase one. Any wave, including time, must consist of three phases, not two.

10.9483.

The new generation of computer processors will be liquid or even gaseous. In liquid systems, the data transfer rate is higher than in solid systems, and in gaseous systems, it is higher than in liquid systems. The atmosphere of planet Earth is the most powerful supercomputer in our galaxy.

10.9550.

Man is very similar to the error of nature. It seems that the system is constantly trying to eliminate this error and destroy the person.

10.9853.

What is the world? Forms, vessels with energy. Energy flows from one vessel to another. The humor is that form is an illusion, a state of self-organization of energy. When the energy gets tired of its form, it flows into another form. Energy can take any form it likes.

10.9900.

The most successful predator on planet Earth is viruses. Every day, viruses kill half of the world's bacterial population, but they do not give up and are born again. The war of viruses and bacteria generates the wave function of life. Man in this struggle is like mushrooms. Mushrooms in this system are the third force that stabilizes and maintains the balance. We can't allow either side to win, if the war ends, it will destroy life.

8.1379.2.

An unfulfilled goal or undone work always cause a serious blow to the karma so that it will be difficult to compensate for it.

8.1466.1. Sociologic chemistry.

From Maslow's hierarchy:

3rd level- need for love,
4th level – need for self-respect
6th level -,need for beauty
All the 7 levels-lust for life, lust for pleasures. Basically, meeting any need brings joy and pleasure,meeting all needs brings absolute pleasure.

0 – system always has zero as a basic source of energy. Zero is pain, it's absence of joy. Originally there is just zero, one should do something to get 1.

8.1466.2. Show with good chemistry.

Turning TV on, I see a lot of movies showing love story of stupid people against a background of beautiful land-scapes. This show has good chemistry. H, B, C, O are taking part, many chemical reactions are taking place, many useful chemical elements are being produced.

B_4C (boron carbide) is one of the firmest essences on Earth, and it's very cheap. I.e if one needs firmness and stability – 4 shares of love to 1 share of stupidity in the show will create a good bulletproof vest.. B_2O_3,boron oxide is necessary for stable burning. And in general this show has a source of energy.

8.1466.3.

Pride I would compare with carbon dioxide, this product of successful life is inevitable and poisonous in its essence.

8.1466.4.

At the 0 level of Maslow's pyramid is a source of energy - hydrogen. It's pain and emptiness. However, its energy is hidden in the desire to escape from pain. In system 0/1, in system black and white there is only pain and desire to avoid its. The energy is just hidden in this desire. The unit is the lust for pleasure in its purest form. All stages of Maslow's pyramid need, realizing, give joy and pleasure, allow to avoid pain. But joy is always white. If you break the balance of colors, the harmony will be broken and again, one way or another, the pain will come.

8.1466.5.

Zero is pain and suffering, Ane is happiness and joy. In essence, Maslow's pyramid includes 7 human needs, 7 ways to happiness, 7 component element of Ane, 7 component elements of the white color, 7 parts of harmony and happiness.

8.1466.7. Since birth there's only pain and the desire to avoid it.

There are different ways to achieve ane. There are 7 rows in Mendeleev's table and 7 notes in musical staff. Obviously, both sociology and psychology has 7 levels of implementation of The Hierarchy of Needs. In essence, there always exist 7 different ways to achieve the same result. Just as in chemistry, life can be built in any row of the table, so as in life when there are 7 ways to achieve a result.

8.1466.8. What is zero and ane in psychology.

Pain is a motivation to avoid pain. If there's no pain, thee's no motivation. The very fact of existence of 1 results from the desire to avoid 0, but when there's no 0, there's no 1.

8.1466.9.

The peculiarity of life existence is such that balance and harmony are consciously shifted in it. That is to say, it was originally created imperfect in order to strive for perfection.
***

If we compare it to a colour chart, we can see that to create a balance, we need either 3 or 7 colours of parts of white plus black. That is, the minimum is black, red, yellow and navy-blue, but life is built on black, green, blue and navy-blue.

8.1472.4. The Universe of zeros.

The question is why make the system 0:1 more complex? Why break down Ane into 3 and 7 parts. Obviously, it is connected with the demand for accuracy of computation. Sometimes it's neither 0 nor 1, in other words the system considers any unclear number a zero. That's why this Universe has more zeros than anes. Many zeros are half-done anes. Ane is quite a perfect construction that has the right to lay claim to energy and time and that's why rigorous selection criteria are applied to it.

8.1472.5. A source of energy.

Zeros are sources of energy for Anes. Ane is a perfect construction that has attained harmony and formed a clear immaculate white color without any tints. It can be said that the closer some number to 1, the more energy 1 gets. The distance from 0 to 1 can be called a level of energetic charge.

8.1478.1. Zombie model "sectarian".

Sectarianism, as a current, originated in India and served as one of the ways to create happy slaves. The main methodology is related to the methods of creating artificial pseudo-happiness, a state of overdose of happiness. Various methods people have turned to happy vegetables. The fact that an overdose of happiness is a big problem, in this state degrades the brain. A person cannot think, but only wants to experience happiness. In fact, this is a type of addiction. Sects are looking for unhappy people who do not know how to live, do not possess philosophy, people who live in eternal pain, and offer them a number of techniques that relieve this pain and give happiness. Basically, it's drug dealers giving a poor patient a shot of heroin.

Methods synthetic happiness: exercising (corny but true),

group therapy (man - gregarious animal in the group, calm and comfortable. Singing, jumping, meditating or doing sports in a group is a great pleasure). There are different methods to ensure victory over him (the man needs a win, win brings happiness). Isolation of oneself, "I am special", "not like others". And "in past lives, I was an outstanding person." It is often suggested that the current life is a dream, and reality is perceived only in a dream or meditation. Goes a powerful blow on 2, 3, 4 levels needs on Maslow. In the second stage, people who do not have their own goals in life, create a variety of pseudo goals and objectives, begin to teach all sorts of nonsense, and slowly output to 5, 6 and 7 levels of realization of needs...

All this causes a rather critical level of happiness in the body, incompatible with normal life or with the work of the brain. In a state of permanent pleasure it is difficult to think critically. In fact, over time, these slaves begin to work for the sect for pleasure. They are constantly recruiting new members, trying to plant important people and leaders, rich people, who then provide powerful support to the organization as sponsors and patrons. And no one asks anyone for anything, just to help people is considered a good tone.

On the first, second and third glance, it's pretty harmless, but in fact, synthetic happiness, like any other drug, one way or another, destroys a person's life, turning him into a slave of pleasure, a slave of love, slave of happiness.

Recently, the technique implemented in the ranks of office workers and salespeople through various technical trainings, etc. of Large employers with very best happy slaves.

By the way, if a sectarian is deprived of another dose of synthetic happiness, he will be overcome by withdrawal, pain and depression. Sectarians are almost unable to live in the real world. Escape from reality does not mean going back.

8.1586.2. Mindsex.

Besides classical sex, children and DNA hidden in chromosomes, there also exists intellectually mental sex. Sometimes, DNA structures of two people are compatible in such an interesting way that it brings them great pleasure to talk to each other. Information that flows between these two DNA, gets mixed and gives birth to very beautiful, useful and unique results. Joint things done by such people can be really productive.

An important consequence of this text is the fact that it's better to deal only with those people whose DNA is compatible with yours. When sex is great, it results into producing beautiful healthy children.

8.1586.2. Thincoitus.

Besides classical sex, children and DNA hidden in chromosomes, there also exists intellectually mental sex. Sometimes, DNA structures of two people are compatible in such an interesting way that it brings them great pleasure to talk to each other. Information that flows between these two DNA, gets mixed and gives birth to very beautiful, useful and unique results. Joint things done by such people can be really productive. An important consequence of this text is the fact that it's better to deal only with those people whose DNA is compatible with yours. When sex is great, it results into producing beautiful healthy children.

8.1789.1. Stick and carrot.

Angel and demon are two-faced Janus. He's an angel with good people and he is a demon with a whip with sinners.

8.1930.4.

Light at night is evil, darkness at daytime is evil.

8.1941.3. Strange conclusion.

Given that human is most similar to the virus by nature, any medicine that kills diseases is poison for him. In particular, the truth is the killer of the virus of foolishness and lies. If all foolishness and lies are killed in a human, he will die. Consequently, human, as a virus variation, fears and hates the truth.

8.1992.1.

Many people don't get the nature of miracles and they think that a miracle is something unbelievable but in fact, a miracle is when things go smoothly. When everything is great and goes according to plan, it's a miracle in itself.

A miracle is when troubles do not happen and your enemies sink into hell even at a mere idea of hurting you.

8.1992.2.

A miracle is when everything goes smoothly.

8.2127.5.

You can only make a mistake not to more extent that you were right.

8.2483.3.

A television set is, in fact, a device that transforms reality; it is a window into a different world, whence something immaterial creeps into our world in an attempt to become material and acquire a frame. Such portals to a different world are, in addition to television sets, the Internet, books, newspapers, magazines as well as other means of transferring and commuting information...

8.2483.4.

A television set is, in fact, a programming unit. The task of this and other similar tools is to program human thoughts

THE THEORY OF EXISTENCE

in the same way as microcontrollers are programmed. A signal encodes thoughts, programs of action and various sets of service commands.

8.3680.1. Spirit of sex.

Speaking of good sex, it can be added that probably the best sex takes place in the Russian smoke bath. The Russian smoke bath is a specific phenomenon, which remains only in remote villages. The smoke bath is mild, it's comfortable to be in, the body steams well, it's possible to have sex right in the steam room. A steamed feminine body opens all its chakras and the feeling is just incomparable. Moreover, it's very comfortable in the bath, there is heat that opens all the body chakras and pores and comfort at the same time. The smoke bath is absolutely natural and living, it literally breathes the fire spirit and this fire gets into every corner of body and soul. In an ordinary classic Russian or Japan bath or Finnish sauna the feelings are completely different, there is too hot, too dry, or not hot enough and there is no chakra opening effect..

I think that the real spirit of the Russian people, people of fire is hidden in the Russian smoke bath... The wild fire Russian spirit.

8.3680.2. The Russian black banya.

The smoke banya is absolutely lively and natural, it literally breathes the spirit of fire and this fire penetrates into every corner of the body and soul. And here, in the Russian smoke banya, the true spirit of the Russian people, the people of fire is concealed... wild fiery Russian spirit.

9.4095.4. elements: earth, air, fire and water.

The human mind is based on three essences: perfection, uniqueness and usefulness, which are the Foundation of its philosophy.

10.10534. The matrix?

What is the real world? A frequency range within which different information is transmitted at different frequencies. The mind in this system is an information transceiver capable of receiving, interpreting, changing, and generating information.

10.10947. Mutual use.

It is curious that the Moon, rotating around the Earth, warms its bowels and generates its volcanic activity. It is believed that the planets of the solar system similarly warm up the Sun.

10.11128.

Our world reminds me of a perfect kindergarten. Many perfect perfect toys, among which children run and learn to live. When the child grows up, he will become a God, but in the meantime, he is allowed to enjoy and play.

10.11189.

Syntalism believes that many of the laws of nature are directed against man, because man by nature is extremely similar to a virus that antivirus software will sooner or later destroy. On the other hand, we are very happy that there are clever ways to circumvent the law, not fall under it, and again... this law will work if you cross the line, but sometimes you can lie low and wait. Viruses have been fighting fungi and bacteria for billions of years, and they are still the most alive... On the third hand, although a person looks like a virus, it would be more correct to define it as a Trinity of viruses, bacteria and fungi.

10.11243.

Place and time are one inseparable entity. The same place in different times is different independent entities.

10.11578. The cycle of energy in nature.

Life is movement in the sense that the system is energetically closed, the amount of total energy in it is a constant, but to live, organisms must eat, that is, consume energy from outside. Life is the movement of energy from a living cell to a living cell. First they get energy, then they give it away, that is, they eat them.

10.11579.

What is the matter? Some stable state of matter. Any matter, in fact, is energy and is capable of conversion to other types of energy.

10.12301.

Reality is the chaos of non-existence, from which everyone extracts the very being that he is looking for.

10.12564.

The outside world is a broadband frequency range where a signal that can be interpreted as truth and reality is repeated at certain intervals at different frequencies. A signal that is false and an illusion fades out of its frequency range.

10.12639.

The idea of an object changes the perception of that object. By being aware of an object in one way or another, you can use it according to your understanding of its nature.

10.12756.

Objects, if you look closely at them, crumble to dust. Ashes are smaller items that, when combined, turn into larger items.

10.12757.

The laws of being determine the relationship between objects. In fact, this world consists of points that are constantly grouped into different forms.

10.12799. DNA of Ideas.

An idea is a virus, a disease that infects reality, changes it, or even tries to kill it. I am glad that having had an idea, reality develops immunity to It. It's frustrating that viruses are constantly evolving. Interestingly, virus DNA makes up up to 30% of the DNA of reality and significantly changes its existence.

10.13040.

A real person is a virus whose purpose is to look for vulnerabilities in the ideal algorithm of being, thus improving being. God is an ideal order that man always struggles with and periodically wins.

10.13062.

Bread and salt is reality. If you increase the salt from 0.7-1% to 4-5%, you will get an energy source. If the salt becomes 90%, you will get a deadly nuclear weapon.

10.13072.

There are many things that are bad, but if you combine them together, they will become good. That's why they say "don't judge", because the judgment highlights the details and then it's hard for an inexperienced eye to put everything back together.

10.13083.

What is virtual reality? A conceivable connection between abstractions that allows you to instantly move within the universe of space/time, as well as outside it, visiting other universes.

10.13114.

I saw God, God is the inner side of the world. God is an infinite number of gears and wheels that make up the mechanism of being. God is the work that creates being. I have seen God, God is bliss. I have seen God, God is infinite beauty.

10.13224.

While studying foreign languages, I noticed that the same letter in different languages can be translated into sounds such as t, C, W, s, h. the Chinese word "tea" in different languages turns into TII, te, cha, Shai, SAI, Tsai, etc.

10.13302.

In the human brain, next to each other, thoughts about objects that are significantly distant from each other in time and distance can instantly arise. A similar picture should be present in the real world. Movement in the universe is possible on the principles of associativity.

10.13431.

Once an event has occurred, it is almost weightless. If this event is repeated thousands of times, it will gain weight.

10.13440.

Having moved to the savanna, the monkey, in order to fend off predators and hunt, was forced to develop a mind that could invent weapons. There's nowhere to run from predators in the savanna. Running prey is also not catch up. We had to invent sticks, spears, traps, bows, etc.

10.13443.

Being is movement. The first movement symbol is fire. If the fire goes out, existence will become non-existence and life will die.

10.13480. Greenhouse effect and cooling.

From too much is always bad. All good this medicine only in small quantities. In large quantities, everything inevitably turns into poison. On the other hand, the bad thing is that it feeds on poison and comes to eat this poison, so reducing its amount.

10.13631. The theory of probability.

There is chance in life, but love, that is, order, manages it quite effectively.

10.13634.

They say time doesn't exist. It's not like that. There is only time, we live in a stream of energy called time. Nothing exists except time. The three-dimensionality and movement of this flow creates space.

10.13727. The order of this movement.

What does God mean by order?
- Rhythm! All processes of being are rhythmic and similar to the beating of the heart. Listen to the rhythm of your heart and you will understand what is fear and joy, life and death, why order is movement, and movement is life.

10.13733.

Divide and conquer. Any discontinuity is the product of a thirst for power, that is, pride. Maximum discontinuity is chaos. There is no passage of time in chaos. The time is right.

10.13734.

Any feeling is continuous, like a continuous space of energy. The discontinuity creates a form, the illusion of the real world.

10.13735. Play, musician.

What you call the material world is rather a musical rhythm, a kind of melody that vibrates the single space of non-existence. Being is the music of non-being.

10.13740.

We live not in a two-phase world, but in a three-phase world, where the past and the future are in opposition to the present.

10.13766. A multi-faceted world.

Reality is very multifaceted. Any, even the most simple and banal essence, can be viewed from many sides and sides.

10.13775.

Coral Paradise is a stupid, slow, but nonetheless the most effective predator on planet Earth.

10.13817. Sperm.

The world is the unity of being and not being. Being is what it is. Non-existence is something that does not exist, but potentially it wants to be and fights for a place under the sun. In fact, nonexistence is a sperm.

10.13865. Good and two extremes.

The worst and the best are always the same. There is no struggle between black and white. There is only the good and the best, where the good is the Golden mean, and the best is the two extremes, one in nature.

10.13878.

Mindfulness is the realization that a grain of sand today and a grain of sand yesterday are two different grains of sand. Every grain of sand is a universe, in the few minutes that you haven't seen it, millions of years could have passed inside this universe, changing everything dramatically.

10.13966.

The problem with a drug is not that it is bad, but that it makes ordinary life bad. The drug turns the good into the bad, and then disappears itself. The output is only bad.

10.14000. Problem.

Vice is too much fun, which is very overheated CPU. A by-product of any Vice is fear. There is an overflow of fear and its sublimation into anger, laziness, attention deficit, procrastination, desire to escape, persecution mania, phobias, depression, contempt, negativism, desire to strangle and ...not enough for you?

10.14024. The Mirror Is Nothing.

Since all judgments are false, the opposite judgment is automatically created for the balance on each judgment, so that the balance of zero is not violated.

10.14025. The uncertainty principle of energy.

The principle of "you – me, I-you" is related to the principles of "mirror of nothing" and the principle of immutability of the energy balance in closed systems. Inside hydraulic systems, the amount of fluid is constant. Energy, space, and time are all liquids.

10.14026. The Joker's Riddle.

The balance of energy in reality is unchanged. However, reality is the unity of being and non-being. In the space of non-existence, you can invent energy. The invented energy of non-existence can be used in reality, provided that the commandment "do not judge"is observed. As long as we don't open the black box, we don't care if Schrodinger's cat is alive or dead, but we can use the cat as a Joker in a real game.

10.14045.

The soul of order is wise and has seen a lot... But she likes to look at the world through the eyes of chaos. The soul of chaos is young and mischievous. The procedure is little in the world of surprises and delights, and the chaos gives one the feeling of colorful life.

10.14119.

Women can be metaphorically called the immune cells of being, which are responsible for fighting viruses, bacteria, and diseased cells. It can also be assumed that the immune system contains information about the RNA of cells. It is the RNA that determines which DNA will be used by the stem cell to find its place in life.

10.14120.

There is an empirical assumption that information about cell RNA is located in the body's immune system. In addition, the immune system is responsible for analyzing (loving) viruses and inserting their DNA into the body's DNA. By studying their enemies, the immune system then defeats them with their own weapons.

10.14122. Chaos is a clock.

What you think is chaos is a clock. Chaos is a huge mechanism where thousands of parts and gears do their work, forming the course of time. To see time in chaos, you need to be careful. Mindfulness is love. Love chaos, and you will become the master of time.

10.14150.

Darkness is a lie. You look at the night sky and see points of light in it, and the darkness between them, but it is not darkness, it is you who are blind. Peering into the darkness, you will see the light of billions of stars in the sky. Any dark point in the sky is just a distant light.

10.14169. Natural despotism.

When we observe being, we see that nature, in creating the laws of nature, is not interested in the opinions of those whom these laws control.

10.14170. Honest law.

All are equal before the laws of nature, even nature itself.

10.14174.

From the point of view of psychology, the soul is a software package, and the brain is an electromagnetic computer system. So you can think of your soul as a virus that has enslaved the poor monkey's brain.

10.14178. The illusion of pain.

From the point of view of neurobiology, fish do not feel pain, they do not have receptors for this and the brain is too primitive. The illusion of pain is unknown to fish.

10.14192. The limit of pleasure.

All your emotional UPS and downs are a metaphor for Sisyphus. Sisyphus is a proud man, hungry for power. Power is a great pleasure. Having reached the top of pleasure, Sisyphus gets used to it and falls to the bottom of suffering. What kind of injustice do you see in this system? If Sisyphus is not cooled, it will simply overheat from pleasure and burn out. Sisyphus is thrown into hell for the most humane reasons.

10.14224.

Vices don't spoil people, but they make them uniquely different. Perfect perfection is a pure white sheet. Every man, with all his vices, has a unique pattern that is of some use if you take the trouble to find it.

10.14253.

To see reality as it is, you need to bring all the details together. The wholeness is the reality.

10.14323.

It's not as simple as it seems. Healthy on the outside is often rotten on the inside. Rotten on the outside, often quite alive inside.

10.14344.

Human pride is the greatest enemy of the ecology of the planet Earth. Pride is a bottomless hunger and thirst for consumption. Pride is a lack of love for the real world and a consumer attitude to life, aimed at getting pleasure at any cost. What does it mean to protect The earth's ecology? This means protecting our familiar and friendly environment. Nature does not have bad weather, nature will survive in any weather, but people will not survive the destruction of their usual ecosystem.

10.14391. Awareness itself.

They say that truth is incomprehensible, but not for those who have love in them, because love is the truth. On the other hand, love is eternal and love is knowledge. When you know love, you will kill it. How to kill an immortal?

10.14446.

I saw the world with different eyes...

10.14508. God is the law.

God exists because God is law and order. There are laws of nature and being, and only a fool would deny it. God is like Santa Claus. At Christmas, you find gifts under a beautiful idol (Christmas tree). It wasn't an idol that gave them to you, it was Santa Claus. Then you accidentally find out that the gifts were put under the tree by your parents, and you start to think that Santa Claus does not exist. You're stupid, think about it, who made your parents put gifts under the tree? Santa Claus is law, order, and ritual.

10.14562. A moth flying into the fire.

You can take all the joy and pleasure you can from life, of course, especially if you are going to live for a very short time. In a short period of time, any lie is possible. However, in the long run, only truth and moderation are stable.

10.14727.

Any activity is useful, even if you were eaten, you were useful as food. The only question is who gets this benefit. The slaves of vices are useful; their time passes through their vices to the masters of vices, who serve higher purposes.

10.14734.

The viscosity of things depends on time. Some things flow into a puddle quickly, others for a long time. It is always interesting to observe things, how long they will last, preserving their integrity.

10.14736. Absolute zero.

There is nothing terrible or very good anywhere, everything has its pros and cons, which, when added up, inevitably turn into zero. Zero is truth, nothing exists but zero, everything else is only an illusion of consciousness.

10.14737. Illusion of the universe.

The existence of the universe is an illusion. There is only one absolute zero, and all other deviations from it in different directions only compensate for each other, leaving the existence of zero unchanged.

10.14740. Cathode and Anode.

Pride reaches for pride to burn in hell. Love reaches out to love, to rejoice in heaven.

10.14741.

Our existence is very similar to a kind of integral unity of the battery and the generator.

10.14852.

Order is a simplified and primitive form of chaos. Be simple and people will reach out to you, because order is a pit.

10.14873. An era of change.

Vices are needed in order to stabilize the system, protecting it from changes when it has reached a certain state that is generally harmonious and acceptable. If the defect is removed, changes will begin. Vice is like sleeping and stopping, getting rid of Vice is like waking up and starting to move.

10.15017. A closed line is a circle.

The entities of shortfall, moderate, and bust are looped in a circle where moderation is white and the rest is black, with shortfall and bust converging into a single straight line. And if you look at this construction from the outside, it is a ball, one side of which is in the light, and the other in the dark, and this ball rotates and flies somewhere.

10.15018. Colored and white nights.

The light condemns the darkness of night and not of a great mind, but from ignorance. It is unknown to light that there is more than enough light in the dark, and many inhabitants of the night are more sensitive to light and see in different ranges of radiation. For the resident of a night as the day, and what do you call the day, it's a dazzling and super bright light.

10.15063. The Mission Of Chaos.

Chaos should not judge ourselves by the standards of the Order. Chain of the order of chaos cause misery. Nevertheless, order is useful for chaos in that it allows you to restrain yourself from falling into extremes. In turn, the mission of Chaos is to save Order from its extremes. The problem with chaos is that it doesn't like what it will die without. Order needs stupidity, spontaneity, and uncertainty, but it is afraid of them. That is why the main mission of Chaos is to save Order from itself.

10.15071. Order never lies.

When order tells chaos what not to do, chaos intuitively wants to do the opposite, because it knows that order is lying. What is true for the order of chaos is a lie.

10.15107.

The meaning of the unity of chaos and order is that stupid things must also be done. Chaos, taking energy from order, must realize the stupid, unprofitable, useless, but beautiful. The meaning of chaos is to create beauty and thus inspire order, infusing it with life.

10.15127. Everyone has someone else's stupidity.

What chaos regards as its love and truth, order regards as stupidity. What order regards as truth will be a direct ticket to hell for chaos.

10.15140.

The form is a vessel of energy. Form is what holds back energy. To get access to energy, you must destroy the form. Pride is the desire for power over energy, which tries to enslave energy and give it form.

10.15192. The same subjects.

The size of items depends on the amount of attention paid to them. If you give a small object all your attention, it will increase it to cosmic proportions. On the other hand, if you take a small and large item, and then pay 100% attention to each of them, it turns out that they are the same size and significance.

10.15196.

The DNA of light is 99% dark. Moreover, light is very similar to RNA.

10.15207.

It is not necessary to argue with the order and customs. Order is love. Love should be admired and tenderly offered various decorations, justifying that it can be beautiful and appropriate. Order is like a woman, you can offer it, but you can't impose it.

10.15262.

The most beautiful thing about our world is that nothing lasts forever.

10.15306.

Uncertainty is a random generator that allows love to live forever, for love is knowledge.

10.15316.

God is the main idea of the universe. God is an idea, a perfect dream, the pursuit of which motivates the movement of life. The dream generates the laws that are required to achieve it. Laws are building algorithms that can be followed to achieve a specific goal.

10.15424.

It is a lie that primates (monkeys) were vegetarians, for

they ate worms, insects, and eggs wonderfully. The idea that people can do without animal protein has no historical roots.

10.15459.

Darkness is a diminution of light. Light is a diminution of darkness. I also thought that what you call light is a rather weak light, 38% percent. After all, if the light is 100%, you will go blind, immersed in solid white.

10.15469. Different frequency ranges.

Death is the slowing down of life, then death accelerates and becomes life again. Life and death are two different intensities of the same wave essence.

10.15570.

The speed of light is not the speed limit. The speed limit is Time. Time is everywhere at the same time.

10.15572.

Be careful, black and white often change places. If you miss this moment, you will lose awareness. A certain thing today is good, but tomorrow it is already bad, then it will become good again, but everything has its time.

10.15583.

The world is dominated by love alone and nothing else. Love is the God and ruler of this world. Well, those who do not have love, they live in another world, which is called hell.

10.15584.

Love is much higher than money. Love is the owner of money, which tells who and where to spend money. Love is a cruel but fair master. Anyone who dares to disobey it will instantly fall into hell. Anyone who obeys it will be happy.

10.15585.

Half of the world is built in projects that are unprofitable, useless, and unprofitable. The reason is love. If love is the ruler of the world, he says, you must do it, then you must do it, and anyone who resists it will suffer.

10.15587.

God is the law of natural selection, according to which the strongest survives. The strongest is the one who has love in him. Love is sensitivity and attention to the outside world. Love is knowledge, courage, courage, honesty, and strength. The one who has love is the strongest. This world was created by the law of Love.

10.15588.

The survivor is the one who adapts most effectively to the external environment around him, in order to extract energy from it and protect himself from external threats. In other words, you will need Love to survive. Love is mindfulness, knowledge, and kindness to one's neighbor. By integrating as much as possible into the environment, a person firmly occupies an ecological niche where he is most useful.

10.15589. Survive useful and beloved.

When a living creature finds and occupies its ecological niche, it can be seen that not only does it choose a place, but the place also chooses it. All living things in their place are useful and perform some very important tasks. To eat, you must allow yourself to be eaten.

10.15594. Vices as a mechanism of specialization.

Curiously, vices allow us to implement the concept of narrow specialization. Having limited himself wherever possible, such a person invariably leaves something in which he attains excellent perfection.

10.15609. Inner strength.

Truth is something that acquires the ability to grow independently from a point to the universe. Having gained inner strength, the point expands its boundaries and gains access to new energy within its boundaries. Having received this energy, it once again expands the boundaries, and so on to infinity. Truth is love.

10.15675.

On the one hand, everyone is different, on the other, the same. The trees are essentially the same, but there are no identical trees. Forests are also all different, although the same.

10.15730.

In fact, man is also a fish only shallow. Of course, he can't live in space, but he feels quite comfortable in the gas environment.

10.15739.

Metaphorically, Chaos generates information and beauty. Order is beauty and order. In fact, Chaos is the mother of order and beauty, and order is a rebellious child who loves and hates his parents.

10.15781.

The mind is a product of pride, so the greater the hunger and the hope to eliminate it, the better the brain works.

10.15804.

Reality is much more interesting and unexpected than any fiction that a person can come up with.

10.15828. Kindred spirits.

We know that son and father, mother and daughter are all people of the same model, but with different settings. Let's say there was originally one or more people, for example, seven. Therefore, there are only seven models of people that are compatible with each other in one way or another. Moreover, there is an opinion that people of the same model are not compatible with each other. Others, on the contrary, say that a soul mate is good.

10.16007.

In solitude, chaos dissipates and dies. Alone the order turns to stone and dies. The Unity of chaos and order is life.

10.16074.

The same road leads to hell and heaven. But the one who is greedily hasty, comes to hell by it, and the one who is restrained and does not hurry, comes to heaven.

10.16089.

Matter is a reflection of an idea, but ideas also reflect the desire of matter for beauty and perfection.

10.16147.

Fears are suffering about what is not there. Your freedom of choice is such that you can decide for yourself whether to believe what is not there or wait for it to appear. However, I noticed that what is not there does not appear immediately, but grows very slowly from a small sprout. And if you pay attention at the very beginning and pull out the evil by the roots, then nothing will grow.

10.16322.

I saw angels in Paradise, they were always happy, creating

the world. I saw demons in hell, howling in hell, exhausted from their sufferings, unable to bear any more pleasures, boredom and boredom. From an overabundance of pleasure, pleasure turned into pain. Alas!.. there was nothing in hell but pleasure, which became pain.

10.16324. The third dream.

Having merged with God, we traveled with him through the universe, enjoying the contemplation of its beauty. An endless journey through an infinite universe. The infinite variety and novelty of beauty did not cause satiety, for nothing was ever repeated.

10.16334. A hive in the world of flowers.

According to the worldview created by syntalism, there are three entities: hell, Paradise, and reality. Paradise is a place where angels joyfully create existence out of nothingness. Sources of joy here are work and moderate portions of honey. The third world is a being where flowers of beauty grow. Angels-bees pollinate flowers, collecting nectar here, then turn it into honey and place it in hell. In hell, there are stores of pure honey and pleasure. There is nothing in hell but a concentrate of pleasure and joy. The nuance is that the honey is obtained more than necessary, and it needs to be disposed of somewhere. This and a number of other tasks are solved with the help of demons (slaves of Vice and pleasure). For the sake of getting honey and ending up in hell, demons are willing to do anything. Bees, of course, protect hell from demons, issuing them honey dosed for work, but demons are greedy and persistent. Those demons who still manage to get into hell, doom themselves to eternal torment. Too much pleasure turns it into pain. An eternity of pain is unbearable.

10.16384.

All the millions of illusions of darkness add up to one darkness. What is not is equal to what is not. All the millions of colors and shades of white also add up to one white, and they are all there, and they are all equal to each other. Thus, any illusion is equal to any reality.

10.16402.

The universe can be represented as a wave system extended in 4 dimensions. A kind of infinite number of growing soap bubbles embedded in one another.

10.16469.

Curiously, black and white is a matter of light intensity. Color diffraction can also be applied to darkness if you increase the sensitivity of the receiver or change its range.

10.16499.

The probabilistic nature of being allows us to learn, because it forgives our imperfections and allows us to accumulate perfection. However, imperfection cannot live long. Those who are untrained quickly drop out of the game.

10.16515.

I didn't see anything useless in nature, so everything that is useless is doomed to destruction, and what is stable is useful in some way.

10.16585. Flash of light.

What is the universe? This is a four-dimensional soap bubble that expands. An infinitely fine line of pure white paper expanding in time.

10.16601.

We are three-dimensional beings living on the surface of a four-dimensional globe. In fact, our universe is a four-dimensional ball, that is, a three-dimensional plane. Wherever you go, you will always come back. Curiously, this ball expands, increasing its volume.

10.16610.

All that is dead contains the seeds of life. This should be known and remembered.

10.16624.

Truth is a Schrodinger cat sitting in a black box. This box cannot be opened, because the truth will then turn into a lie.

10.16629.

Nudity is not quite natural, primitive people were covered with hair.

10.16639.

God is a reference generic object, all other entities have differences from it, but it (God) includes. In fact, the system accumulates information about differences from the standard.

10.16649.

The Trinity of the world is that it consists of three elements. Reality is a garden of beauty, where bees collect the pollen of joy, turning it into honey (a concentrate of pleasures). In another world, this beauty is created. And in the third world, honey is stored. In fact, the third world is hell, the second is heaven, and the first is the real world.

10.16675.

The universe is not a ball, but a ray like a cone. Moving in space time, the universe forms a cone shape. In fact, it is a point source of light, which, as it moves away from its source, scatters and fades.

10.16676.

If we assume that the universe is a point source of light, then our reality is like a ray of light. However, the light shines in different directions, so there are parallel intersecting time processes with a displacement along the entire diameter of the original ball.

10.16685.

A lie rules the world for a moment, but at this moment its power is absolute. For a moment, a lie is powerful and useful. At this moment, the truth is harmful and weak. However, when it comes to years, the truth becomes useful and good, and the lie is harmful and weak. A lie is an impulse that fades in time, and the truth is an average straight line.

10.16686.

The wave is the essence of the unity of lies and truth, momentum and the moving average, something that endlessly dies and is reborn, obeying a single aspiration.

10.16694.

Progress is infinite, because knowledge is infinite and truth is incomprehensible.

10.16695.

There are a lot of stable States of ones and zeros, they are all extremely close to each other. If something doesn't work, we change it a little bit, and it starts working. To break, respectively, everything is just as simple.

10.16714.

Perfection is a pure white leaf, but it is no better than a pure green leaf or a red one. Getting closer to white doesn't make the color any better, either. Each unique color is equal to another.

10.16766.

Why is blue worse than green? We know that white thinks he is perfect, and we will not even argue with this, but does the proximity of green to white make it better than blue?

10.16773.

They say mistakes lie on the surface, but the truth is deep. Does this mean that the shape of items and the differences between them are determined by the accumulation of errors in the system? The whole essence of beauty is errors that have created a difference from the standard.

10.16790.

Reality, reflected in the illusion, is modified and returned back to reality, changing it.

10.16796.

Stupidity is the negation of the utility of something. Everything is needed for something. Everything performs some useful functions. The system is scalable, if something disappears, something else will start performing these functions.

10.16803.

The past and the future are two extremes that strive to connect in the present to one point.

10.16804.

A thing now and a thing in the future are two extremes, one of which tends to become the other.

10.16812.

Things grow in time from one extreme to the other.

10.16830.

The speed of light is small, the speed of thought is much faster. Lies and illusions are always in a hurry and always late.

10.16835.

God is the absolute, the unity of black and white, and life is a movement between these two extremes.

10.16836.

Energy tends to accumulate information about all its possible forms.

10.16839.

Life is a struggle for creation. Creators fight among themselves for energy and the ability to create. Whose ideas are more beautiful, whose love is stronger, he wins this war for a place in the sun.

10.16840.

Only what is unworthy of life perishes, everything that is worthy of life, if you want to kill it, you will not kill it. Truth is as tenacious as life itself

10.16844.

Any default force includes a self-destruct mechanism. The stronger a person is, the faster they will self-destruct.

10.16851.

To see God, you need to descend from the heavens created by your own pride, bend down, take a grain of sand in your hand, look at it carefully and say: "Hello, God!»

10.16866.

The tree doesn't look very much like its seed, but it did give birth to its seed. So God is a point, and the universe is a tree.

10.16885.

Circumstances are the main law of existence.

10.16914.

The absolute is love, beauty, truth, perfection, but there is no limit to perfection. Perfection is an infinite array of various States that depend on circumstances. Perfection is like a mountain range, where not one mountain is an idol, but infinitely many, like the stars in the sky.

10.16922.

The absolute is not one big star, but the universe. The power of the absolute is in the infinity of the starry sky, and not in the size of individual stars that think they are idols. There are asteroids, comets, planets, moons, stars small and large, black holes. Black holes are stars that have reached the limit of their mass and burned out, folded into a point. All these are parts of the absolute. The star does not approach the absolute, it is just a part of it. The absolute accumulates information about all its possible States.

10.16925.

The absolute is a virtual matrix of time and space. Each space-time cell has its own piece of the absolute. And time and space are an illusion, an information pointer to a memory cell.

10.16981. The absolute is a matrix.

The idea of an unambiguous absolute is the idea of devilish pride. The absolute is the infinity of diversity. Ignorance is to think of white as a single entity. White is an infinite variety of colors, their intensity and interaction options.

10.16982.

Black is moderate white. Increasing energy turns black into white.

10.16983. The light of truth.

Modesty is not about light, light takes everything that its energy is enough for.

10.17011.

Space is four-dimensional, time is the fourth dimension of space. The universe is a four-dimensional matrix.

10.17040.

Of course, nature is fighting with the man. Nature is the truth, and man is a lie that brazenly pretends to be the truth.

10.17056.

All living things come from the same clay.

10.17057. A perfect mind.

They say that the human mind is the crown of evolution, the strongest of the strongest. This is a delusion, the crown of evolution is God.

10.17320. Quantum superiority.

True randomness is a very rare thing. For example, the human brain is not capable of randomness. Only quantum computing systems can generate random numbers.

10.17321.

Even in the real world, randomness is a huge rarity that still needs to be caught.

10.17327.

There is no limit to perfection and there is no imperfect.

Complex perfection is a set of simple perfections. And what you call imperfection is not enough perfection.

10.17350.

True black. True randomness. Is an imperfection. These are all very rare and therefore valuable items.

10.17354. The hardness of the world.

Everything is good. Chaos is a lot of order. Perfection is the multitude of imperfections. Light is concentrated darkness. Order turns darkness into light. Darkness has the property of permeability. The light is persistent.

10.17493.

There are no items, only a process. Forms are like the flow of water in an ocean of energy, like an image on a TV screen, like a virtual reality world.

10.17514.

This world has no purpose, it moves from nowhere to nowhere. However, the world will be happy if you come up with goals for it, and even help you implement them.

10.17517.

Black doesn't exist, but white doesn't exist. White is a balance of other colors.

10.17526.

Zero is a metaphor for things that have a form. The guardian of borders is called Mr. Devil. The devil makes sure that no one breaks the law and does not cross borders.

10.17569.

Any of the options for future events is equivalent. It is a delusion to think that something has more chances or less.

The probabilities of all events are the same. Any of the 10 options will happen with a 10% probability.

10.17624.

Light is the initial form of hardness. When light grows, it becomes solid and material. The entire material world is the evolution of light.

10.17667.

Reality is the chaos of love. Love is order. Thousands of layers of order are superimposed on one another. Every lover in the chaos of existence sees his own love, his own ideal order.

10.17709.

The stars also Shine during the day, but you are blinded by the sun and can't see them. The sun creates a white noise in which weak light sources are lost.

10.17720. Humility with life

What is the Kingdom of the dead and death? This is a kind of virtual world where an extinct race of gods immigrated. Life is a Paradise where the dead are periodically released to enjoy reality. It is a great honor to live in the real world. The infinity of illusions is unbearable, the dead are already sick of death, they are ready to live on any conditions.

10.17769.

The growth of the entropy of chaos is not the destruction of the system, but its striving for perfection. The growth of chaos is the growth of layers of order in the system. The order strives for power and control. The more order, the more chaos. Restraint of order reduces the level of chaos in the system.

10.17788.

Our world is a hell in which there is nothing but beauty and pleasure. Too much good causes pain and suffering. To save the situation, you need to come up with a job and a dream, come up with fear, come up with suffering...

10.17789. X Factor.

The idea that the world is perfect and I'm not is stupid. The world is a perfect hell, full of joy and pleasure... an imperfect person is just necessary to him as salt to bread.

10.18508.

Nothing came up light. Light gave birth to time; time consists of space. The slow light is converted into matter.

10.18868.

A point is an entity outside of time. The universe is a point distributed in time.

10.19313.

The world is absolutely just, because justice is one of the names of order. If you look closely at the real world, you will notice that it is absolutely ordered.

10.19315. Step into the abyss.

What you call the destruction of the system and the growth of the entropy of chaos is an amazing thing when the system itself grows. In fact, we are seeing a fall into the abyss, a system of units begins to divide within itself and create new forms. The avalanche-like explosive growth of the number of these forms seems to you to be chaos, but, in fact, it is many different layers of order. The greater the depth of separation of forms in the system, the more such a system generates energy.

10.19635.

The future is a stream of unstructured energy raining down into the present, where there are various ordered systems that catch this stream and convert it into other forms of energy.

10.19663.

Death is not the end, but renewal. The real world is something like an immortal multicellular being, where there is a constant process of updating living cells.

10.19690.

Light is a wave process, but waves are a kind of illusion. In fact, a wave is a particle that is constantly destroyed and re-created. This is the nature of light, this is the nature of love, time, and life in General. Every moment our world is destroyed and created again.

10.19749.

The universe expands not in space, but in time, where time is a stream of unstructured energy. Space is structured energy.

10.19774.

Reality (present) it resembles a pancake in a frying pan, where the future is the flow of energy that fries this pancake.

10.19806.

Things are separated from each other in time and space.

10.19850.

In order for a system to generate time, it must be divided into three dimensions. Growth and movement in three dimensions generates a flow of energy in the fourth dimension. The fourth dimension is time, the simplest form of energy that can be expressed as money or matter.

10.20344.

Love is the power that gives shape to things. In the absence of love, objects lose their shape and energy turns into chaos.

10.20367.

There is no struggle for a place in the sun. Point already exists. In the future, accumulating information and reaching a critical mass, the point begins to decay inward, so turning into the universe.

10.20721.

Time is what gives shape to things, but time is also what releases energy from the chains of form.

10.20722.

Time is the primary information flow. Information is what gives shape to energy, thus creating the real world.

10.20757.

I see three essences of being ... energy, information, and (gender)a guide between them. The guide is love. Information is beauty. Energy is joy.

10.20867. Goals give birth to means.

We cannot say that man is a Creator, but man is a creation... A reasonable man is a creature of gratitude, grace, and nobility. Man is a tool that creates. Just as man creates an instrument to achieve his goals, so God created man to achieve his goals. A person is a means that creates other means by which certain ideas are realized. I believe that an idea, wishing to realize itself, creates a whole chain of production facilities, one of which is a person.

10.20883.

Man is not the crown of evolution as such, but a fairly perfect tool with which nature has discovered a new way of creating new and unique forms. Those forms of the real world that are generated by man are a new evolutionary step in the development of the universe.

10.20903. Spoiled food.

Parasites are very useful for their victims, because they protect them from other predators and parasites. Symbiotic parasites kill their victims for a long time, allowing them to reproduce and live. However, getting into foreign organisms that do not have protection from this parasite, they quickly destroy it. Thus, extraneous predators avoid eating other people's food, feeling a threat to their own lives.

10.20989. Eight-dimensional space.

The real truth is the whole from many different perspectives. In addition to our three standard dimensions, there is also a measurement of time and potential. Potential is another dimension, directly inaccessible in our three-dimensional reality, but expressed through the future and randomness.

10.21037.

The hardness of this restrained movement. Matter is very slow motion.

10.21100.

The world is a hydraulic system where transitions create kinks.

10.21233. Eraser.

Let there be light! The creation of light was an act of destruction of darkness, the further process of disintegration of light is an act of creation of darkness. Perfection is darkness. That is, in fact, the creation of the world began with taking an eraser and erasing a part of nothingness to make room for a new creation.

10.21451.

In this world, everything is a constant, the existence of variables is a subjective illusion of the mind.

10.21494.

The world is very different. People are different. As many people as there are worlds. It all depends on the paradigm of what beliefs to look at it through.

10.21525.

I think that information and energy are the same thing, only from different sides. The more information you have, the more energy. The more energy, the more information. There is no information without energy and energy without information. Only the forms in which the energy resides change. Money is the simplest form of energy. Things and people are more complex forms. The more perfect the form, the more energy is concentrated in it.

10.22213. Negative world.

We live in a negative world, because darkness is perfection, that is, information that gives form to matter. Light is pure energy, devoid of forms. Light is a stream of energy, time, the future, non-existence.

10.22358.

The proportion of salt and fresh water on the planet Earth is very similar to bread with salt. And the proportions of land and water are similar to the Golden ratio.

10.22496.

The acceleration of an object's speed creates a time difference between the inside of the object and the outside. The higher the speed, the faster the outside time speed is. At the same time, the speed of time inside an object can remain the same. In fact, it is not so much about increasing the speed, as about illustrating the thesis-lovers do not notice the clock. You think it's been a couple of days, but it's been decades.

8.1447.11.

Any matter is both bad and good at the same time. It's you who should decide whether count 0 or 1. With all else being equal, nothing may stop you from capturing only pluses of any situation.

ABOUT THE AUTHOR

8.2479.

SoloINC (anc.greek "combining the uncombinable", keeper of the grain")

Soloinc Logic, philosopher from the city of Sofia. Soloinc (Diamond Solo / Solodilov Dmitry), Bulgarian psychologist and Stoic philosopher. Supporter of the merger of logical and sensory methods of cognition. He considers the connection of traditional philosophies with modern science. He is the founder of the cyberphilosophy of Syntalism (Quantum Nanophilosphy), which considers the problems of philosophy, sociology, psychology and economics in terms of systemic cybernetics and logic.

Soloinc is not the first, but the last philosopher. Evangelist and cyberpunk guru. The author of more than 73 thousand original ideas and thoughts. Main books: "Variothoughts", "Diamond Stoic", "Theory of Existence", "Money Bible", "Quantum Philosophy", "Mathematics and Progression", "Velerechie", "The Device of the Mind", "Royal Buffoon", "Liberastia" , "Surrotic", "Surfutur" and others, in total more than 888 books.

3.1753.

In fact, Variothoughts is very tedious. I have seeked the truth all my life, then I found it and concealed it in a different place. Variothoughts is an intellectual quest and a mosaic of truth, broken into thousands of pieces. I found the truth in plain sight and concealed it back as well as before... What's the point? It's a game or a way to kill boredom. We live eternally and boredom turns our life into hell. I want to save you from sufferings for some reason...

10.21128. Soloinc Music

Soloinc Music is a stunningly beautiful integrity of music and text, admiring metaphors and secret meanings. Soloinc Music is a pleasure for living minds who have dedicated their lives to the search for beauty and truth. Soloinc Music awakens the minds and ignites the heart. Everyone will find joy and strength to live in it.

10.2341. A realistic mysticism.

The genre of poetry and music of Soloinc is a mystical realism. Most Soloinc songs are mystical ballads or religious hymns, prophecies, and insights. Soloinc lyrics are always metaphors and mystical signs. They cannot be taken literally. These are grains of sand in which entire worlds are hidden. All words are the opposite. To understand the meaning of the Variothoughts texts you need to read from bottom to top, from right to left.

SYNTALISM - GENERATIVE QUANTUM NANOPHILOSPHY

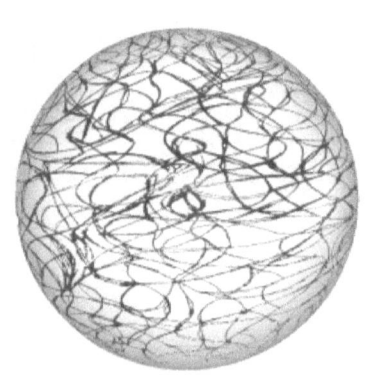 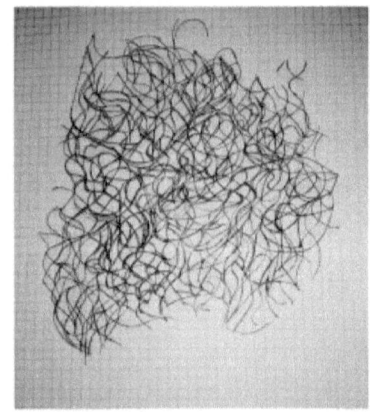

10.19296.

The philosophy of Syntalism was inspired by the poetry of life, expressed in the poems of such poets as Shakespeare, Robert Burns, Williams Blake, Pasternak, Lermontov, Mayakovsky, Velimir Khlebnikov, Paul Eluard, Andrey Bely, Alexander Blok, Voznesensky, Asadov, Gutseriev, Anna Akhmatova, Tsvetaeva, and others. Where if the philosopher had not come, the poet would have been there. Poets are like rays of light showing the way to thinkers.

5.782. Syntalism is the philosophy of the 5G generation.

Small thoughts are the philosophical system built in the millimeter wave range. Syntalism is 5G philosophy in the millimeter wave range built according to generative genetic algorithms.

5.767.

In Variothoughts, conceptualization follows the generative genetic algorithm.

5.768.

Variothoughts is the self-teaching guide on generative philosophy.

10.22348.

Syntalism is a philosophy that connects the unconnected with the goal of achieving integrity. Integrity is truth. To know the truth, the mind must cultivate tolerance and humility.

5.783.

Variothoughts is structured as a phased antenna array that ensures a dynamic horizontal and vertical growth of thought according to the generative algorithm and makes it possible to create different-sized logic data arrays. This solution minimizes energy consumed to maintain the integral information field. Variothoughts is a system of small cells in the millimeter wave (super-small thought) range in which the size of cells and their interaction structure are dynamic in nature.

10.3109. Unified system of knowledge.

The philosophy of Syntalism is by far the most perfect and

clear philosophy, revealing the nature of being. Syntalism is like an ocean containing all other philosophies and religions. Syntalism understands and explains any point of view, agrees with everyone and loves everyone, considers everyone beautiful. Thousands of points of view, uniting into streams and rivers, turn into an ocean of Synthism.

VARIOTHOUGHTS COLLECTIBLE BOOKS

4.3423. Sand vs truth?

A book's collectable from the Variothoughts series costs only 1 cubic meter of real estate property. It is a very delicious price for something priceless.

3153.

God loves collectors as they give work to many creators...

6.6033.

The electronic version of Variothoughts is huge but printed versions are more complete and this book's collectables and handmade versions are unique in their completeness. Each of the author's gift manuscripts of Variothoughts is handmade and customized, that's why it includes even the latest texts that exist only in rough copies and have not yet been published anywhere.

10.22513.

Friends, I have not sold any Variothoughts collectibles yet. Pride rules people, that is cowardice and greed. There are very few courageous and intelligent people. He who is brave and buys the first book is very lucky. The first collector's copy of Variothoughts is a great value. Each collection book is registered and numbered. However, there will never be many of them, if I sell such books at least a few pieces a year, it will be good.